SAN FRANCISCO PUBLIC LIBRARY
3 1223 90061 7300

S0-BFP-294

Microbial Contamination Control Facilities

Microbial Contamination Control Facilities

Sponsored by the

Biological Contamination Control Committee
American Association for Contamination Control

Edited by

ROBERT S. RUNKLE

Administrative Manager

and

G. BRIGGS PHILLIPS, PH.D.

Director
Becton, Dickinson Research Center
Becton, Dickinson and Company
Rutherford, New Jersey

VAN NOSTRAND REINHOLD COMPANY

New York Cincinnati Toronto London Melbourne

Van Nostrand Reinhold Company Regional Offices:
Cincinnati, New York, Chicago, Millbrae, Dallas

Van Nostrand Reinhold Company Foreign Offices:
London, Toronto, Melbourne

Library of Congress Catalog Card Number 70–81349

Published by VAN NOSTRAND REINHOLD COMPANY
450 West 33rd Street, New York, N.Y. 10001

Published simultaneously in Canada by
D. VAN NOSTRAND COMPANY (*Canada*), *Ltd.*

15 14 13 12 11 10 9 8 7 6 5 4 3 2 1

Van Nostrand Reinhold Environmental Engineering Series

MALVERN F. OBRECHT, Ph.D., P.E., *Consulting Editor*

Thermal Insulation, by John F. Malloy

Industrial Waste Disposal, edited by Richard D. Ross

Microbial Contamination Control Facilities, edited by Robert S. Runkle and G. Briggs Phillips

Acoustics, Sound Control, Vibration Isolation, by Lyle F. Yerges

Van Nostrand Reinhold Environmental Engineering Series

THE VAN NOSTRAND REINHOLD ENVIRONMENTAL ENGINEERING SERIES is dedicated to the presentation of current and vital information relative to the engineering aspects of controlling man's physical environment. Systems and subsystems available to exercise control of both the indoor and outdoor environment continue to become more sophisticated and to involve a number of engineering disciplines. The aim of the series is to provide books which, though often concerned with the life cycle—design, installation, and operation and maintenance—of a specific system or subsystem, are complementary when viewed in their relationship to the total environment.

Books in the Van Nostrand Reinhold Environmental Engineering Series include ones concerned with the engineering of mechanical systems designed (1) to control the environment within structures, including those in which manufacturing processes are carried out, (2) to control the exterior environment through control of waste products expelled by inhabitants of structures and from manufacturing processes. The series will include books on heating, air conditioning and ventilation, control of air and water pollution, control of the acoustic environment, sanitary engineering and waste disposal, illumination, and piping systems for transporting media of all kinds.

MALVERN F. OBRECHT, PH.D., P.E.

Consulting Editor

Acknowledgments

The American Association for Contamination Control expresses its gratitude to the following individuals who served as members of the project team of the Biological Contamination Control Committee responsible for developing the initial documents from which this book has developed:

Everett Hanel, Jr.	*U.S. Army Biological Laboratories* *Fort Detrick, Frederick, Maryland*
Raymond J. Dugal and J. J. O'Rourke	*The Ralph M. Parsons Company* *Los Angeles, California*
Tom B. Lanahan	*S. Blickman, Inc.* *Weehawken, New Jersey*
Edward Rich	*National Institutes of Health* *Bethesda, Maryland*
Paul Maupin	*Pitman-Moore Research Center,* *Dow Chemical Company* *Zionsville, Indiana 46077*
Jack F. Zanks	*Langley Research Center, NASA* *Hampton, Virginia 23365*

The above individuals prepared sections or portions of sections based on their respective experience and training for the first draft of this book. In addition, they have continually served as technical consultants and reviewers during the final editing of the manuscript.

We gratefully acknowledge the remaining members of the Biological Committee who reviewed and criticized the book at various stages of its development. Without their contribution of knowledge, time, and effort, this undertaking could not have been completed.

W. Emmett Barkley	*National Cancer Institute* *National Institutes of Health* *Bethesda, Maryland*
Robert W. Edwards	*Enviro Systems, Inc.* *Dulles International Airport* *Washington, D.C.*

Robert K. Hoffman	*Physical Defense Division* *U.S. Army Biological Laboratories* *Fort Detrick, Frederick, Maryland*
Martin G. Koesterer	*General Electric Company* *Valley Forge Space Center* *Philadelphia, Pennsylvania*
John R. Puleo	*Spacecraft Bioassay Labs* *U.S. Public Health Service* *National Communicable Disease Center* *Cape Canaveral, Florida*
Albert P. Kretz, Jr.	*Wilmot Castle Company* *Rochester, New York*
Joseph J. McDade, Ph.D.	*Dow Chemical Company* *Bethesda, Maryland*
Dick K. Riemensnider	*Environ Services Section* *Division Hospital/Medical Facilities* *Silver Spring, Maryland*
John A. Robertsen, Ph.D.	*Director Clinical Lab Research* *Bio-Dynamics, Inc.* *Indianapolis, Indiana*
Donald Vesley, Ph.D.	*University of Minnesota* *School of Public Health* *Minneapolis, Minnesota*

The editors would like to express their own appreciation to the American Association for Contamination Control, its officers, and to the Executive Secretary, Mr. William Maloney. The support which has been given to the Biological Contamination Control Committee during the past few years, and to the editors during the preparation of this book, is sincerely appreciated.

The gratitude and respect of the senior editor is extended to his wife, Betsy Runkle, who has had to "live" with this book for the past years. Without her understanding, support, and encouragement, the book could not have been completed. The support of his previous employer, the National Cancer Institute, National Institutes of Health, is also gratefully acknowledged by the senior editor.

Robert S. Runkle
G. Briggs Phillips

Rutherford, New Jersey
April, 1969

Contents

CHAPTER 1

Introduction

This book attempts to correlate factors, concepts, and information that influence the planning, design, construction, acceptance, and operation of facilities wherein the control of the microbial elements is required. Control of microbiological contamination has increasingly become a criterion for many diverse areas in our society. Technology in the handling of micro-organisms infectious to man has undergone revolutionary changes during the past twenty years. Laboratories serving the medical, public health, and veterinary professions have played an increasingly important role in man's struggle to cope with infectious diseases. These laboratories perform diagnostic services, produce vaccines, develop chemotherapeutic agents, operate in the area of national defense, serve as teaching centers, and are the instrument of the epidemiologist in controlling diseases in the population.[98] Since the start of this country's efforts in the exploration of space, the concept of contamination control, both biological and physical, has become increasingly important and demanding. Although the initial problems facing NASA and the numerous contractors in the aerospace program concerned development of equipment for construction and assembly of components and devices in a dustfree environment, recently the problem of maintenance of other planets as an ecologic preserve and the prevention of possible back contamination from other planets has added a new dimension and technique to the space effort. Another area that traditionally has been aware of the problem of contamination control is the drug and pharmaceutical industry, both in the safe production, packaging, and distribution of drugs and pharmaceuticals and in the manufacture and assembly of sterile disposable medical items. An exotic example of the application of contamination control in the area of patient care has been the use of closed environments surrounding a patient bed, known as the Life Island,* which provides a germfree environment for patients whose resistance[113] has been lowered, such as during intensive chemotherapy[112] or treatment of

*Life Island ®—Patient Isolation Unit, Mathews Research Inc., Falls Church, Va.

burn patients.[57] More recently, attempts are being made to utilize laminar flow rooms or beds as a substitute for the Life Island in patient care[130] or the surgical suite.[40]

It is recognized that many documents exist that deal with one or more aspects of this subject. However, none of these is all-inclusive and many have not received wide distribution. It is hoped that this book will be helpful as a *reference* document to management personnel and operating staff embarking upon a new facility program, to the architect/engineering firms charged with the responsibility of design, to biological and engineering safety personnel responsible for facility acceptance, and to the biohazard program responsible for safe operations. Several recent documents have been prepared that should serve as additional reference materials and that contain more detailed information on specific subjects.[33,35,58,101,108,118,135]

A microbiological barrier, as described and discussed in this book, is a device or system that will prevent or limit the passage or migration of microbiological contaminants. These barriers can be primary barriers that immediately surround the hazardous procedure or the item or product to be protected, or secondary barriers that surround the primary barrier and provide additional assurance of the containment or exclusion of microbial elements. Figure 1 depicts both primary and secondary barriers. Since many documents have covered the subject of primary barriers,[14,50] this book will deal primarily with the secondary barriers for microbial contamination

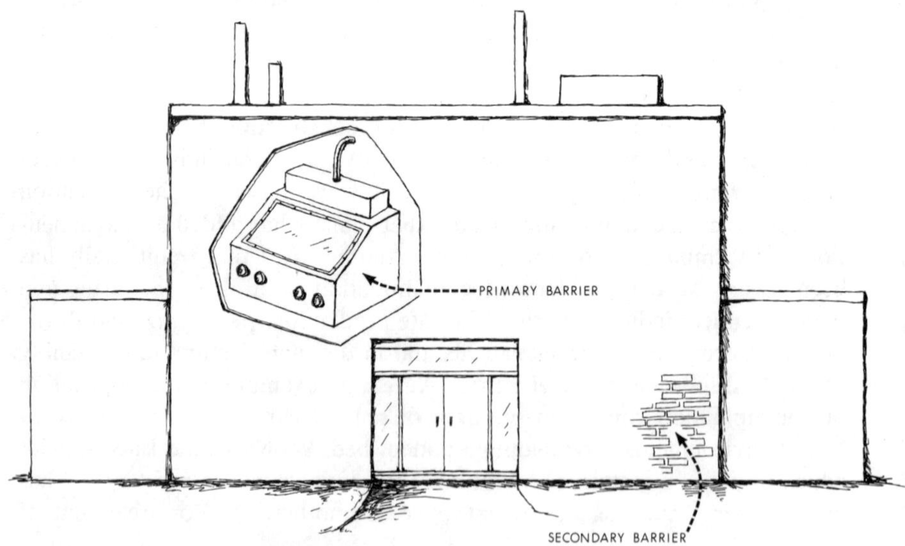

Figure 1. Primary and secondary barrier concept.

control and will present methods for their incorporation into new or remodeled facilities.

BACKGROUND

The early development of techniques and methods for microbiological isolation and mechanical barrier devices[19] can be viewed as a first attempt at contamination control. Bacteriological barriers such as the flasks of Schulze, in 1836, and Schwann, in 1837, were employed to test the theory of spontaneous generation and heterogenesis. The chamber that Tyndall used in 1868 to show the relationship between the light-scattering ability of aerosols and the ability of airborne organisms to initiate growth in various infusions was an example of an early microbiological barrier. In 1861, the obstetrician Semmelweis recommended a hand-washing technique for the control of sepsis in childbirth. This was followed, in 1867, by Lord Lister's treatise on antiseptic technique that gradually evolved into what we now know as aseptic technique. The laboratory isolation apparatus used by Davaine, in 1870, Lister, in 1878, Koch, in 1881, Petri, in 1887, and others enabled the development of the pure culture techniques that put the science of bacteriology on a sound footing. All, essentially, were biological barrier devices.

Classical methods of contamination control are mainly concerned with techniques and employ only a minimum of equipment. They depend heavily upon hand operations in which the training and skill of the worker are used to prevent sterile objects from coming in contact with contaminated materials. In light of our present knowledge of the ease of the creation and spread of microbiological aerosols, it is not surprising that these classical methods do not suffice for more exacting contamination control needs. One of the first examples of a completely closed and sealed system for biological contamination control was the food canning process invented by Appert and explained in his book, "L'art de conserver pendant plusieurs années toutes les substances animales et végétales," published in Paris in 1810. Placing foodstuffs in a tight container and heating to destroy all microorganisms could preserve the food for long periods of time.

Starting in about 1885, scientists interested in studying life isolated and apart from the influence of microorganisms and other contaminants developed a number of mechanical barrier and isolation systems. These researchers started by copying the apparatus of the early gas chemists. Some of the earliest biological barriers were nothing more than the flasks and bell jars that had been used by the plant physiologists during the controversy over the means by which nitrogen is fixed into plant tissue. One such apparatus was the 4-liter jug used by Berthelot[10] in an attempt to grow plants on sterile soil. Eleven years later Nuttall and Thierfelder[88] published the results of their experiments with bacteria-free animals using a modified bell jar. Other modified bell jar barrier systems have been used by Cohendy,[22] Balzam,[6] and others.

In discussing microbiological barrier apparatus, Trexler[124] states that ". . . a closed system to be of general use must be so designed that the contents can be observed, manipulated and materials introduced and removed without destroying the mechanical barrier." One of the most elaborate early cabinets to satisfy these criteria was devised by Kuster[67] for rearing germfree goats. Kuster's cabinet was the first to use arm-length rubber gloves. It also contained essentially all the features of modern-day isolators for germfree animals, including an entrance air lock and air supply filters, and was operated at a positive air pressure. This apparatus was improved by later workers such as Glimstedt[45] and Reyniers.[106] Today a variety of mechanical barrier apparatus is used for research with germfree animals. These have been reviewed by Luckey.[71] The predominant types are the heavy-walled stainless steel isolator of Reyniers,[106] the thin-walled stainless steel tank of Gustafsson,[52] the flexible plastic isolator of Trexler,[123] and the rigid plastic isolators of A. W. Phillips.[90]

Microbiologists handling infectious disease agents have long realized the need for mechanical barriers to internalize hazardous procedures. Safety cabinets were in use in German laboratories early in the century.[43] Shepard, May, and Topping[115] at the National Institutes of Health developed a wooden cabinet for hazardous laboratory operations. In England, Van den Ende[128] developed similar cabinets for use during large-scale production of scrub typhus vaccine. The first stainless steel microbiological barriers for infectious disease work were described by Wedum.[132] Other types and improvements have been described by Reitman and Wedum,[105] Phillips, et al.,[93] Gremillion,[50] Blickman and Lanahan,[14] Wedum and Phillips,[135] and Phillips.[99]

CURRENT ASPECTS OF MICROBIOLOGICAL CONTAMINATION CONTROL

It has been said that the task of science is to understand nature. But in trying to understand and in contemplating nature we have all been impressed perhaps with the almost inconceivable dichotomies we observe. It taxes our imagination, for example, to try to fathom how immense nature can be. The size of our galaxy, the existence of hundreds of thousands of millions of other galaxies, and the concept of galactic distance expressed as the distance that light can travel in 10, 100, or even 1,000 years is almost beyond our ability to grasp. On the other hand the minute aspects of nature become just as impossible for us to fully appreciate and understand. In relation to our concept of atoms and molecules and in thinking about the physics of atomic structures we might wonder if some relationship exists between an electron and a nucleus of an atom and a sun with its orbiting planets.

Likewise we reflect, as we observe nature, on the relative size of *living* things. The dichotomy here can be represented by the gigantic size of living structures such as the General Sherman Redwood Tree on the one hand and, on the other hand, viral particles that are as small as 10 millimicrons in diameter. Truly, the scope of nature is breath-taking.

In relation to the diminutive side of nature, as science and technology have more and more understood and made use of the properties of smaller and smaller things and have undertaken tasks where smaller and smaller entities can interfere with operations, the need for controlling contamination that might interfere with essential processes has become a reality. While it is perhaps possible to conceive of contamination control in a "large" or on a planetary scale (a meteorite as an earth contaminant), we think primarily today of contaminants that are relatively small in size.

This book deals with an interface between microbiology and engineering and in a general manner with aspects of biological contamination control as they apply in a number of areas of current application.

From an ecological point of view the concept of microorganisms as undesirable contaminants is at first difficult to accept. If one based his conclusion on the relative abundance of life of different types and the complicated interactions and symbiotic relationships involved, one might conclude that macroorganisms, not microorganisms, are the contaminants. Life on this planet exists in a literal ocean of microorganisms. There is hardly a niche that does not exhibit this form of plant life. Microorganisms are universal in nature. They are found in water, air, dust, soil, nonprocessed foods, skin, hair, and the intestinal tract and the solid and fluid body wastes of vertebrates and invertebrates. They proliferate in such diverse regions as hot sulfur springs and barren desert soils, in ice deposits in arctic regions, in deep recesses of the oceans and in the upper part of our terrestrial atmosphere. All of us are continually exposed to microorganisms although we may not have been exposed to the fundamentals of microorganisms.

Under normal circumstances there are myriads of symbiotic and other relationships between macro- and microorganisms. The vast majority of microorganisms are not only harmless but play essential roles in the balance of terrestrial nature as we know it. Microorganisms, for example, function in decay processes; they convert insoluble protein to soluble protein; they function with higher plants in the fixation of nitrogen and convert sugars to alcohols.

In addition to the above, there are a few characteristics of microorganisms that are helpful to know in relation to biological contamination control. First, the types of biological contaminants of greatest concern to us are those which are known as bacteria, yeasts, fungi, viruses, and rickettsiae. These unicellular or subcellular forms of life exist in the general size range of 10 millimicrons to several millimeters; most are below two microns in diameter. Being living things, microorganisms reproduce, survive under some adverse conditions, and adapt to live under various environmental conditions. Some secrete a sticky envelope or capsule that will help to protect the cell from the effects of disinfectants or other adverse substances. Others secrete exotoxins which are toxic to higher animals. Some rod-shaped bacteria form spores in their cytoplasm. These are freed into the surrounding environment when the vegetative cell dies. These spores are externally resistant to chemicals, heat, and drying, and some will withstand boiling water. Microorganisms also easily become airborne. They are resistant to cold temperatures. They can utilize many diverse substances for growth and energy. Some require complex organic material such as peptides and polypeptides while others can convert simple inorganic salts into energy and cellular material. The density of most microorganisms such as viruses and tissue culture cells is about 1.10 as compared to a density of 1.0 for water and 1.33 for protein. Finally, an important fact about microorganisms is that "infection," when it occurs, will often depend upon the state of the host and the site of the host-microorganism interaction as well as the characteristics of the microorganism.

The world of living things differs from the physical world in that there is easier recognition of the constant development and change taking place in the former. There is a tremendous variety of microorganisms on the earth, many of which have not been described. These microorganisms utilize as energy sources many very diverse substances ranging from basic elements such as iron and sulfur to the most complex of organic compounds. Some microbial species utilize atmospheric oxygen. Some typical characteristics of microorganisms that are important in contamination control are shown below:

—10 millimicrons to 10 microns
—Ubiquitous in nature
—Adapt to environmental influences

—Some form spores
—Aerobic and anaerobic
—Utilize various substrates
—Resistant to cold
—Easily become airborne

Definitions

Microbiological Contamination. Microbiological contamination is defined as the presence of living microorganisms in a specified environment. In general, these are bacteria, fungi, viruses, or rickettsiae.

Microbiological Contamination Control. Microbiological contamination control is the total procedure that achieves and maintains a desired microbiological state in, on, or around a stated environment, object, or thing.

The overriding philosophy of control of the microorganisms in any system is related to the ability to define the microbial load in the system at any particular point in time. Contamination control is achieved if the microbial load does not exceed the level established as the lowest acceptable limit. Maintenance of control, however, is complicated by the fact that, in opposition to inert contamination, the microorganisms in a population may be going through simultaneous processes of multiplication and death. Insofar as these processes are concerned, the only condition of microbiological control that can be considered stable is that of sterility—the absence of all viable microorganisms.

Man in the Contamination Control System

The significance of man in any system where microbiological contamination control is attempted deserves special consideration and understanding. Because man is an extremely prolific source of microorganisms, and because he can be extremely susceptible to those microbes that are pathogenic for him, inclusion of man in any system usually signals the weakest point in the contamination control effort. Unlike inanimate objects, man cannot be conveniently separated from his microbial flora, and only with difficulty can he be enclosed in a ventilated garment to separate him from a controlled environment. These facts suggest that microbiological contamination control is simplified when man is excluded from the system and carries on his functions in relation to the system in a more or less remote manner.

Much confusion and misinformation exists relative to "people-generated" contamination, both viable and nonviable. While the figures that have been published relative to "personnel emissions" may be indicative of the number of particulates associated with humans in occupied areas, much of such data have not been verified on a scientific basis. Even worse, statements that have been made relative to a proportion of the total airborne particulate count being made up of viable particles are grossly inaccurate and misleading.

What we know with certainty about man in a system that is being controlled for microbiological contaminants is shown below:

—Man intimately associated with microorganisms
—Different skin areas carry different numbers of microbes
—Each individual is different—and the daily microbial profile may fluctuate
—No large differences due to sex or climate
—Airborne emissions of microorganisms vary over wide ranges
—Coughs and sneezes are prolific sources of air contaminants
—Airborne survival is complex

Horizons for Biological Contamination Control

While biological contamination control is at least as old as man's first attempts to prevent spoilage in his food supplies and to isolate persons with communicable diseases, the parameters of interest and concern for the control of biological material have perhaps expanded more in the past decade than in all previous recorded time.

A partial list that illustrates the diverse areas wherein biological contamination control techniques are used includes (1) maintenance of other planets as ecologic preserves (e.g., spacecraft sterilization), (2) prevention of possible back contamination from other planets, (3) protection of hospital patients during and after operative procedures, (4) isolation of burned patients and patients under treatment who are uniquely susceptible to environmental microbes, (5) protection of research workers handling infectious disease agents, (6) protection of researchers handling oncogenic viruses during studies on the etiology of leukemia and other cancers, (7) the same production, packaging, and distribution of drugs and pharmaceuticals, (8) controlled handling of animals used in research in order to prevent the spread of zoonotic disease, and (9) safe shipment and transportation of infectious microbial cultures and medical specimens.

It seems clear, therefore, that the need for biological controls is increasing and that as our society assumes a more complex posture the level at which contamination control applies becomes smaller and smaller. It is equally clear that the engineering disciplines will play a major role in solving these control problems.

With regard to the specific developments in biological contamination control technology that have taken place in the past decade, no one would be so bold as to suggest that the means for solution of all current problems is at hand. On the other hand, the research ingenuity and finesse being applied by engineers and by medical and biological people in the solving of control problems has resulted in a literal revolution in methodology. Moreover, an amazing amount of cross-fertilization has occurred. For example, the techniques of the pharmaceutical industry have been helpful in studying the problems of spacecraft sterilization, while the procedures for rearing germfree animals have found application for certain hospital situations. Perhaps the single most recognized development, however, has been the application of the techniques of the aerodynamicist in providing minimum turbulence air flow streams (laminar flow) for contamination control operations.

With regard to current problems of biological contamination control, and specifically with regard to laminar flow systems, more definitive systems evaluations are needed to provide an adequate basis for maximum application of various types of equipment for the solution of control problems. Use of the basic concepts of experimental methods applied in a strict sense will do much to help in this regard. We should gradually build up a pool of research data and evaluations that will provide a broad basis of judgment for problems of design of biological contamination control facilities.

When we begin to reach this plateau in our development of contamination control, we will be in the best position to pursue cost economy measures with the maximum efficiency. Stated very simply, with many biological contamination control problems the decision that has the greatest impact on degree of containment and operational costs is that which selects an absolute or partial barrier system. For many current applications, use of laminar flow in partial barriers has increased the efficiency and relative degree of containment to a level hardly thought possible a few years ago. Yet, when the contamination control criteria are severe, as when sterility is to be preserved, no substitute for the absolute system is known. In terms of initial cost and operational time, absolute containment systems are almost always more expensive. On the other hand, proper selection of the design factors covered in this volume with relation to engineering feasibility, cost, risk of contamination, and other factors should allow the best control at the best price.

CHAPTER 2

Planning Stages for Facility Design

INTRODUCTION

Comprehensive planning and programming of facilities for microbial contamination control requires early planning action by staff and management to assure that the magnitude of the design effort, acquisition, and operational costs are recognized as soon as possible.[83] These actions are those that are essential to conventional facility planning, but they are of even greater importance for facilities requiring control of microbial elements. This is primarily because such facilities can be expected to have unconventional design requirements and unique operational methods.

Planning a facility for microbiological contamination control likewise requires the execution of a number of basic policy decisions that will directly affect the design methods used for achieving facility objectives and mission.[135] In order that policy decisions can be established for applicable control techniques, methods, and procedures, the contamination control problem must be recognized and evaluated in relation to the facility, its intended operations, and the designated time periods. Planning should include detailed step-by-step analyses of the contamination control problems with full examination of facility requirements as related to intended operations, personnel, equipment, maintenance, and risks.

Planning, of necessity, should provide for early establishment of the lines of organizational communications. The formulation of teams that will function through the acquisition phases must include representatives of the ultimate operating staff since it will be through their participation that the transition to an operational mode can most efficiently be made. By this rationale, the economy and technical continuity for definition of operating practices are achieved. It is anticipated that costly changes can be avoided when the operating staff are an integral part of the planning, design review, inspection, and activation team.

There are many tasks to be accomplished in the planning of a facility for microbial contamination control. The early coordination of these tasks is essential so that the related facets of design, construction, shakedown, and start-up can be recognized and a full consideration of the interaction between each facet can be obtained. A summary network of this concept is shown in Figure 2.

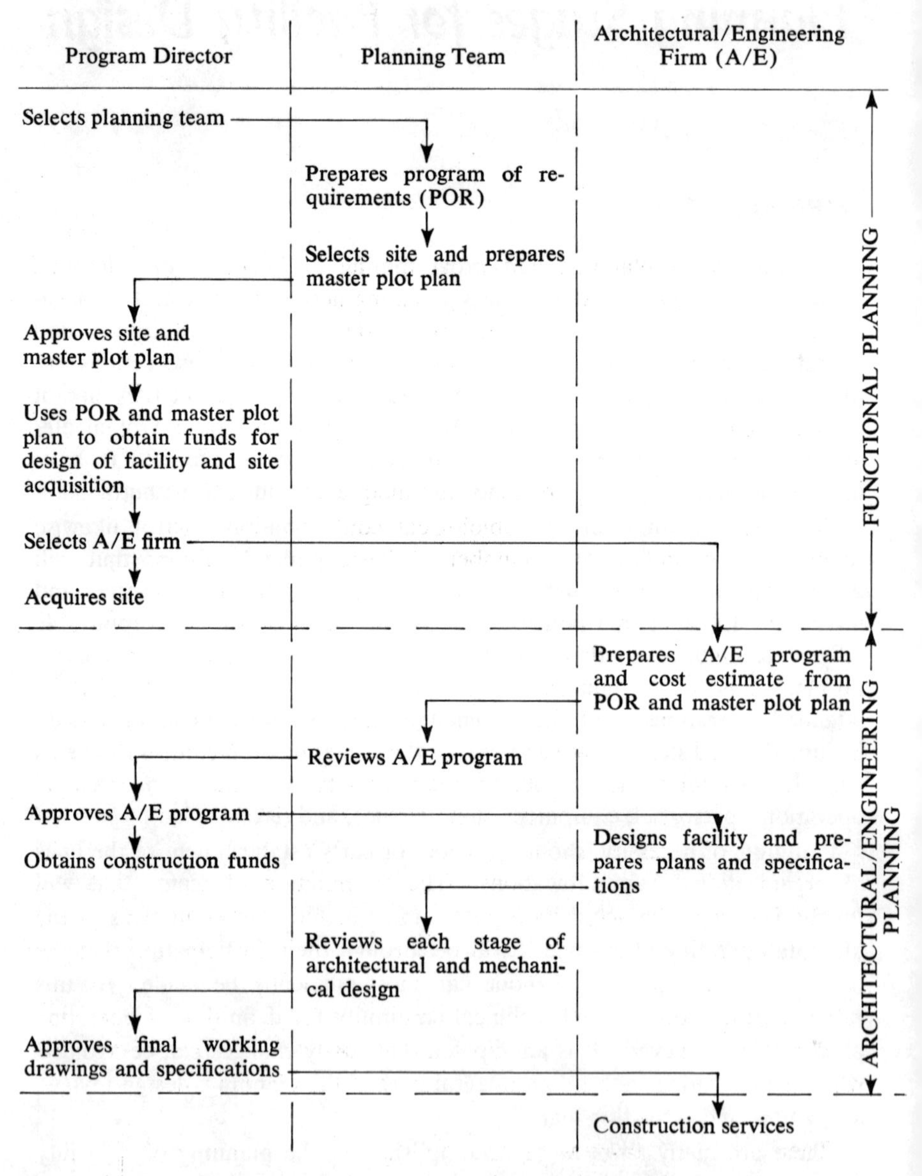

Figure 2. Outline of planning responsibilities.

An integrated approach to the planning of microbial contamination control facilities is presented in the following paragraphs. This approach is directed to all aspects of the design of a contamination control facility. Therefore, it should be emphasized that selective judgment is required in order to use these guidelines most effectively. For a small facility or for some projects involving facility renovation, only those applicable guidelines should be utilized in the planning.

FACILITY DEFINITION REQUIREMENTS

Facility Mission and Objectives

One of the primary planning requirements is that of developing a comprehensive Facility Program Document (or Program of Requirements) for the common guidance of all participants. Prior to the development of such a document, complete descriptions of the mission and objectives of the facility and the scope of operations (relative to types of process or operations) that will be in the facility must be prepared. This information should be used to establish a milestone schedule of the facility acquisition functions.

The Facility Program Document should contain a thorough identification of the total contamination control problem, the specialized control criteria, and the objectives of the contamination control endeavor. This information will facilitate evaluation of their effects on the facility mission and permit development of facility requirements. For example, in the case of a facility for the bioclean assembly of a spacecraft, the first determinations require establishment of the degree of microorganism removal required and a definition of acceptable biological and physical contamination levels. From a facility engineering standpoint, the methods for achieving compatible, feasible, and maintainable decontamination levels or sterilization may prove highly complex or costly. This fact merely serves to emphasize the need to recognize these factors as early as possible. The definition of the control problem should include identification of all known facets, including movement, time, and other parameters, effects, and constraints imposed on facility operations. This step is an essential planning element.

Another example to consider might be the design of a water pollution control facility. Prior to the development of such a facility and associated systems for a community, the physical, chemical, and biological characteristics of sewage and receiving waters and the total process leading to a polluted condition must be identified. The Facility Program Document should include the system for collection and conveyance of sewage and stormwater, methods of treatment, and discharge processes and quantities. The neutral implications of short-term storage of water and the long-range effects of waste discharge on the quality of receiving waters likewise must be recognized.

Analytically derived information needed to develop the Facility Program Document includes:

(1) What is to be done in the facility?
(2) Who or what will be doing it?
(3) What equipment and environment are needed?
(4) What are the site requirements and considerations?
(5) What will the site and community problems be because of operations, hazards, community constraints, utilities availability, etc.?

The following early planning questions by Wedum and Phillips,[135] which relate to the necessary policy decisions mentioned earlier, are representative of the type of basic data that should be developed for successful design and adequate microbial contamination control. These questions were formulated specifically for the design of a laboratory building to handle

Figure 3. Gastight cabinet system. (*Courtesy S. Blickman, Inc.*)

highly infectious microorganisms. They therefore would not be applicable in their entirety to facilities for other purposes. The questions, however, illustrate the thought process which must be followed in the evaluation of any microbial contamination control situation.

(1) Is this building for the use of one man or for a specific project, and will the building at the departure of the man or conclusion of the project be remodeled to suit the next occupant or next project?

(2) To what extent do the views of the facility manager or of whoever has the final authority to determine the level of contamination control design, equipment, and techniques reflect the probable view of his eventual successor?

(3) How many persons will work in the controlled area of the facility, and what will be the conditions of supervision? The larger the number of nonprofessional personnel, the more desirable it is to provide building design and equipment engineered to insure use of the pre-

Figure 3 *(Continued)*

ferred method—from which few deviations will occur because it is easier to do things the right (safe) way. Conversely, the smaller the number of persons, the better the judgment based on education and experience, and the closer the supervision of the group, the fewer the mechanical safeguarding systems required. However, there are some operations in which no amount of judgment and experience can substitute for special microbial contamination control equipment.

(4) What will be the ratio of men to women in the facility? The numbers usually are not the same. More flexibility in personnel policy may be facilitated by dividing the total change room or locker room space in a 40/60 or 30/70 ratio so that the predominant sex may use the larger room.

(5) Does the stated justification and objective of the facility require it to be suitable for study of any microorganism in any kind of experiment, with only the size of the equipment or animal as limitation? If, so, a gastight cabinet system (Figure 3 or equivalent) will be mandatory for some operations.

(6) Will infectious microbial agents be studied as aerosols? Special air-tight chambers and an associated gastight cabinet system are required for many agents.

(7) Will any limitation be placed upon the types of etiologic agents that may be studied? Categories of infectious disease agents and examples of associated recommended requirements for equipment for infectious disease research laboratories are outlined in Table 1. A detailed discussion of cabinets, cabinet systems, and other containment equipment is presented in later chapters.

TABLE 1. Correlation of Estimation of Risk with Recommendations for Protective Cabinets

	Cabinet System[a]	*Single Cabinets*[b]	
Disease or Agent	*Aerosol Studies*	*Aerosol Studies*	*Other Techniques*
Brucellosis	+++		+++
Coccidioidomycosis	+++		+++
Russian s-s encephalitis	+++		+++
Tuberculosis	+++		+++
Monkey B virus	+++		++
Glanders	++	+++	+++
Melioidosis	++	+++	+++
Rift Valley fever	++	+++	+++

TABLE 1.—Continued

Disease or Agent	*Cabinet System*[a]	*Single Cabinets*[b]	
	Aerosol Studies	*Aerosol Studies*	*Other Techniques*
Encephalitides, various		+++	++
Psittacosis	++	+++	++
Rocky Mt. spotted fever	++	+++	++
Q fever	++	+++	++
Typhus	++	+++	++
Tularemia	++	+++	++
Tularemia[c]		++	+
Venezuelan encephalitis'		+++	+
Anthrax	+++		+−
Botulism[c]	++	+++	+−
Histoplasmosis		+++	+−
Leptospirosis		+++	+−
Plague	+++		+−
Poliomyelitis	+++		+−
Rabies	+++		+−
Smallpox[c]	+++		+−
Typhoid		+++	0
Adeno, entero, viruses		++	+−
Diphtheria[c]		++	0
Fungi, various		++	0
Influenza		+	+−
Meningococcus		++	0
Pneumococcus		++	0
Streptococcus		++	0
Tetanus[c]		++	0
Vaccinia[c]		++	0
Yellow fever[c]		++	0
Salmonellosis		+	+−
Shigellosis		+	+−
Infectious hepatitis			+−
Newcastle virus		+	0

+++ = mandatory. ++ = strongly advised. + = optional but in absence of a cabinet a few infections will occur. +− = depending upon technique and supervision. 0 = not required.

[a]Figure 4 or equivalent.

[b]Figure 5 or equivalent.

[c]For persons receiving live vaccine or toxoid.

(8) What methods of animal inoculation will be permissible?

(a) Respiratory challenge: whole-body exposure, head only, nose and mouth only? For these techniques, a protective cabinet and other housing (Figures 3 and/or 4, and Table 1) are essential for the aerosol apparatus and the animals. Each step in the handling of animals and cages must be thought out carefully.

(b) Intranasal, intratracheal, intraperitoneal, intravenous, subcutaneous, intramuscular, intracerebral, oral, etc.? If personnel are vaccinated, it may be possible to perform injections on an open bench top without cabinets, but the range of permissible operations and agents is extended by presence of a protective cabinet (Table 1).

(9) Which of the following animals will be used: mouse, rat, hamster, guinea pig, ferret, monkey, chimpanzee, fowl, cat, dog? Animal caging arrangements must be examined for the possibility that cross-infection between animals may impair the integrity of the experiments. Whether this will happen depends upon the agent, the animal, and the method of inoculation.

(10) Will animals as large as swine, sheep, burros, or calves be used? Rooms for them require good drainage. Flushing-type drains are preferable because of the volume of excreta.

Figure 4. Single safety cabinet. (*Courtesy S. Blickman, Inc.*)

(11) What will be the usual physical form of the infectious material—wet or dry? Dry infectious materials aerosolize more easily than wet and therefore are more dangerous. Maximum precautions are necessary.

(12) Will infected anthropods be grown and studied in transmission experiments? Will any of these anthropods be exotic to the geographic area of the laboratory? Exotic vectors require careful control even if uninfected because of their potential ability to set up an unknown cycle of transmission of disease.

(13) Will there be work with several tissue culture cell lines, used to grow viruses? A special room or enclosure with filtered air may be necessary. Positive air pressure in a room may be requested for uninfected tissue cultures if no clean space is provided outside the infectious unit.

(14) Will large numbers of eggs be used as culture material? Will egg contents be pooled, and into about what liquid volume? Egg trays are difficult to sterilize except by autoclaving. Eggs externally contaminated during inoculation require special precautions during handling, incubation, and subsequent processing. If infection of man has a serious outcome, it is best to have the egg incubator sealed to and part of a gastight chamber wherein the eggs are inoculated.

(15) Is it desirable to be able to change the size, shape, and purpose of the rooms and of their installed equipment from time to time as the years go by? A common finding is that the space needed for animals is underestimated, but the reverse also occurs. Planning for alternative use as laboratory or animal room is worthwhile.

(16) Will the nature of the experiments, the species of animals used, significant change in type of agent and experiment, resident microbial contamination in the room endangering the validity of the experiment and product, nonspecific animal infection such as epidemic diarrhea of mice, potential infection of personnel, or periodic repair or modification by engineering personnel make it desirable that all, or some, rooms, air ducts, air filter plenums, and air filters be sterilized periodically by gas such as beta-propiolactone or steam and formaldehyde? If so, attention must be given to airtightness of walls, ceilings, light fixtures, air duct and utility insertions, windows, if any, and sometimes even the electrical conduit and electrical switches. These precautions also will assist in controlling condensation and vermin. All surface finishes must be evaluated for chemical resistance to decontaminating agents.

(17) Will shaking machines holding microbial cultures be operated in walk-in incubators or refrigerators? In case of breakage of flasks on the shakers there is needed (a) a light switch, an ultraviolet light

switch, and a power switch for the shaker, all located outside the incubator or refrigerator, (b) a view glass in the door for observation before entrance, and (c) an ultraviolet fixture inside the incubator or refrigerator to reduce airborne microbial contamination before entrance after an accident. (In a laboratory, the 10 to 12 air changes per hour will be an effective substitute.)

(18) Will any of the experiments result in animal excreta, the uncontrolled disposal of which would endanger domestic, farm, or feral animals? Examples: anthrax, glanders, equine encephalitides. In general, it must be presumed that all excreta from experimental animals are infective until proved otherwise. Are there any other reasons of law, public relations, volume of material, or microbial virulence that make disinfection or sterilization of sewage necessary?

(19) What is the personnel policy regarding occupational health?

(a) As a condition of employment, must employees accept vaccination with commercial standard vaccines and with experimental vaccines when, in the opinion of the laboratory director, administration of these would decrease the chance of clinically apparent illness? A "yes" answer will reduce the need for mechanical protection, if only those agents are studied for which a vaccine is available.

(b) What level of occupational infection of personnel in the facility is acceptable to management? Subclinical infection detectable only serologically? Minor discomfort no more than from a reactive avirulent living vaccine, which causes only a minority to cease work for one to three days? A "no" answer to these questions may make the difference between installation of individual protective cabinets (Figure 4) and installation of much more expensive gastight systems (Figure 3), depending again upon the agent and the experiments.

(20) For public relations, economic, legal, or other reasons, to what extent is protection from infection of persons not working in this laboratory building considered to be of comparatively great importance?

(21) What are the local zoning laws and building codes with regard to an infectious laboratory facility? What changes in these laws and codes can be foreseen? In what direction are they moving as concerns disposal of potentially infectious wastes and noncombustible trash?

(22) Is study contemplated, now or in the future, of diseases for which permission, and in some cases inspection, may be required by the U. S. Department of Agriculture and/or the U. S. Public Health Service? Government agencies will give great weight to the competence and integrity of the scientist as well as to the quantity of material being handled in determining whether or not a laboratory is

suitable under their regulations. Since these are no clear-cut criteria, if there is any plan to handle highly infectious material, it is strongly suggested that the responsible government agencies be consulted regarding the plans before commitment for construction is made.

(23) Will any equipment, significantly contaminated by an infectious microorganism or toxin, need to be sent periodically to the manufacturer for repair or adjustment, or need to be repaired or adjusted by his servicemen? If so, some arrangement for decontamination of this apparatus may be desirable. For delicate apparatus, ethylene oxide gas is very useful. For economy, it must be used in a leakproof space such as an autoclave.

(24) In the geographical area concerned, are there sufficient dusts, bacterial spores, fungi, molds, or insects so that intake air should pass through a coarse filter?

(25) Where will the experimental animals be obtained? Unless the effort is relatively small, it is better for animals to be produced and prepared apart from the area of research, to avoid complicating accidental infection. A clean area where animals can be quarantined for a suitable time before use is helpful.

(26) As part of a periodic cleaning or disinfection process, will it be necessary to flood the floors? If so, special attention must be given to prevent cracks around floor or sink drains and other locations where monolithic floor surfaces have been penetrated.

(27) How reliable is the source of power for the building? The major danger to employees is an air flow stoppage in a protective ventilated cabinet during a hazardous experiment with aerosolized microorganisms. Perishable refrigerated materials usually can be transferred to a cooler that uses solid carbon dioxide. Contaminated sewage is not a problem if personnel leave the building and the retention sewage tank is large enough for batch chemical sterilization. Inasmuch as work stops, exhaust air is no concern. However, animals in closed mechanically ventilated cages will die in about an hour if there is no ventilation. For this reason, the extent of use of ventilated cages will be a factor in evaluating the need for standby auxiliary power.

(28) Is the probable cost of the laboratory facility fully realized by management? For instance, consider a research building designed so that there are no restrictions in regard to species of microorganisms or their physical states (wet or dry), types of experiments (including use of aerosols), and animals (up to and including the size of a chimpanzee or burro), or combination thereof. A review of a group of modern infectious disease research facilities reveals that only about

43% of the roofed floor space of a typical building will be usable as working space for laboratories, incubators, refrigerators, animals, dish washing, and cage washing. The other 57% will be used for offices, conference rooms, storage, corridors, change rooms, air locks, machine rooms, pipe chases, stairwells, elevators, walls, basement, and attic. Attic space is defined as any part of the attic with at least 5 ft of headroom. On this basis, if the entire building cost is divided by the square feet of working space as defined above, the cost of working space will range from $76 to $147 per sq ft without installed or portable equipment, and from $98 to $179 with all equipment. Pieces of equipment costing less than $200 and the cost of land are not included. Moreover, the cost of building maintenance is high. For example, in one survey, maintenance costs per square foot for laboratory buildings were approximately twice that for residential or apartments and more than four times that for warehouses.

These costs are mentioned because, with the complex design features of an adequate building for research with infectious disease agents it should be made clear at the outset that maintenance costs are of a different order of magnitude from those for some other types of construction. It is recommended that provision be made so that these costs are not charged to the budget for animals, scientific equipment, and laboratory supplies, lest it bear the burden of maintenance costs to the detriment of the research effort.

The preceding questions are indicative of how policy decisions in the planning of a facility for microbial contamination control are influenced by careful analysis and a well defined problem evaluation. In development of the program, answers to the following questions should be available to the planners in evaluating the overall program for making decisions and establishing policy:

—Is the Facility Program Document in agreement with organizational objectives?
—Is the facility defined in the document feasible and will it achieve the intended objectives?
—What are the predictable consequences or risks?
—Has the mission been translated into facility functions and requirements for design engineering?
—Has a literature search been made to determine what has and is being done elsewhere? This search is needed to take advantage of established processes, techniques, or criteria and current state-of-the-art developments.

The Facility Program Document should also identify the need for any additional research effort necessary to further define the problems, to evaluate candidate solutions to the problems, or to verify specific concepts. These staff analyses should be used to delineate the prime and support functions for equipment and personnel.

Development of flow sheets indicating sequence and man/machine allocations of the major operational and maintenance functions of the facility will be valuable for determining the critical and noncritical facility operations and systems, and the interrelationships of these. Since "critical" systems are most likely to influence the cost and design/construction time factors, this segregation is important at the outset of the program to place proper design emphasis on them.

"Critical" systems are those that:
—function in direct support of facility operations
—have special industrial, biological, or radiological safety considerations
—require special testing or development
—have technical complexities which present operation and maintenance problems
—have quantitative reliability requirements
—create mission risks if not operable

"Noncritical" systems are those that:
—provide indirect support of operations
—are static and fixed
—have standard support and reliability considerations

To amplify the above considerations, it may be explained that a facility system (power, environmental systems, etc.) that affects the microbial contamination control operation is considered critical when failure can result in damage or contamination of equipment, the process, or product, or injury to personnel. Each critical facility system must be identified, not only to determine its full technical requirements, but also to establish quantitative reliability for performance of equipment and components within the system. This latter aspect should be expanded in the design phase in terms of engineering, performance, specifications, and identity of support requirements.

Achieving reliability in critical systems is not necessarily accomplished by redundancy of components or systems. The process of achieving reliability where it is a performance requirement or is essential to operation is shown in Figure 5. This process begins with an exposition of system requirements during the early planning of system requirements.

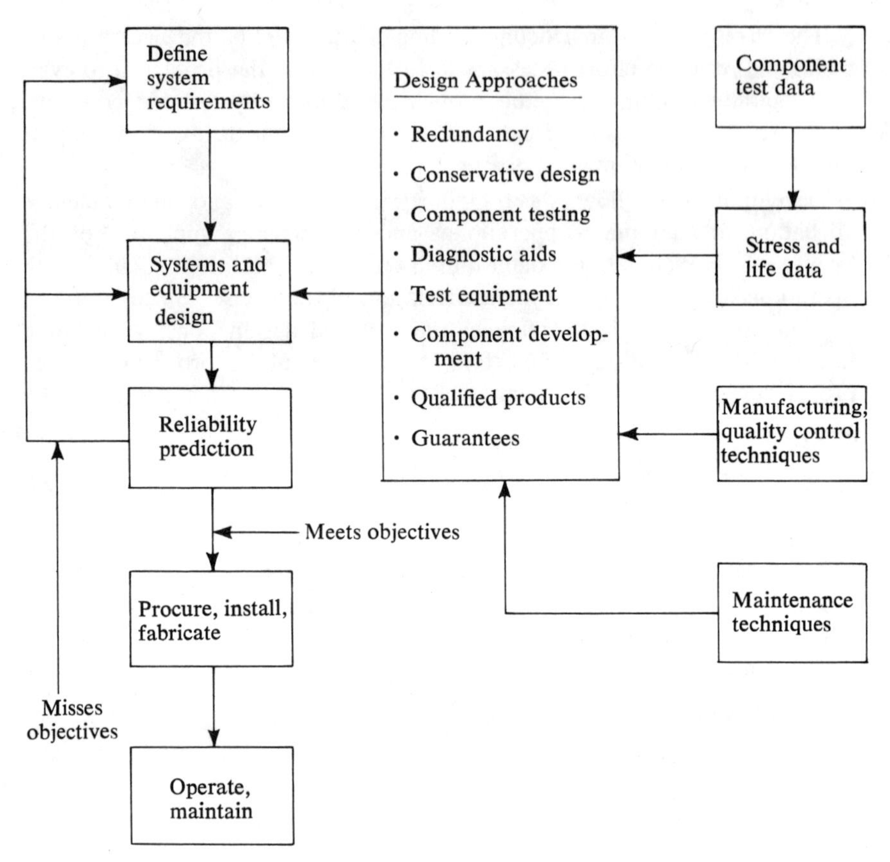

Figure 5. Process of attaining reliability for a system.

Risk of Operations

In the determination of microbial contamination control requirements in relation to facility design, potential hazards and risks should be evaluated on a continuing, consistent, logical, and defensible basis. This may require the architect engineer and the user to perform what may appear to be a thankless and difficult task. It may also require the expenditure of a considerable amount of time in a library without tangible results. Nonetheless, it is recommended that such a risk evaluation not be delayed. In some instances, the architect engineer or user may wish to use a biohazards consultant in such a risk analysis.

The risks will vary depending on the contaminating microorganisms in use or the degree of process or product protection required. The higher the degree of possible risks, the greater should be the emphasis on the control methods and techniques, personnel training, process/manufacturing control, equipment selection, reliability, and maintenance. The "object at risk" may be the product, the personnel, and/or the facility environment, internal and external. Identification of the risks will pinpoint critical operations and systems and provide guidelines to the facility engineer for development of the technical requirements for the next planning phase. The various contamination control techniques and methods selected for isolating personnel from operations or equipment are interrelated to selection of required control modes.

Contamination Control Criteria

As part of the early planning, it is important to provide details and justification for the desired extent of biological contamination control. This decision should not be based solely on concern for the psysiological or constraining impact of microbiological containment on the product or personnel, but on the harmful, damaging, or unwanted effects to the products, processing equipment, or those related indirectly to the product, i.e., personnel, the facility itself, and the physical community environment.

Analyses should be conducted to assure that the following factors are firmly established:

(1) Objective evidence that the unwanted effects are caused or could be caused by the presence and/or actions of microorganisms.
(2) A description of the "contamination control system."
(3) The accurate determination of which point, or area, in the "contamination control system" is critical and verification that the undesirable effects can be eliminated when appropriate control procedures are inserted.
(4) A definition of the total contamination control problem must include the following information related to the overall facility requirements:
    (a) *Microorganism(s) Concerned*
        —number of types; species, genus, family, etc.
        —concentration or density
        —physical form
        —identification of availability of controls
        —requirements of microorganism viability, if any
        —environmental and metabolic needs
    (b) *Mechanisms of Contamination*
        An analysis should be prepared to describe the modes by which microbial contamination may enter and spread through a facility. This analysis typically should provide a theoretical flow pattern of the contaminant, starting at its source and continuing to the

eventual place where the harmful effects are sustained or the susceptible host or object affected. The analysis should allow identification of required contamination control measures.

Facility Siting Requirements

The site itself can be a significant cost factor; however, site costs should not ordinarily be weighted above technological factors during the site selection analyses. It is most important to determine for each potential site whether the proposed facility can be constructed thereon at a reasonable cost within acceptable limits of safety and with adequate operational flexibility. The required environmental characteristics of the site and surrounding regions may identify problems that deal with the physical, operations control, and engineering factors of the plant, and define important facility requirements.

Site analysis includes a compilation, analysis, and interpretation of a variety of interrelated factors. These factors may include such geological aspects as air patterns, water conditions, and meteorology of the surrounding area. Long-range development plans of the area and surrounding territory should be considered. Potential hazards to the community in the event of an accidental release of biological materials must be visualized. An analysis should include such variables as time, area and method of release, concentration of material, etc. Only through such a complete analysis can site selection be optimized.

Site evaluation for one or more potential sites and preliminary data for biological facility features should be conducted as a step in determining the best overall site for the proposed installation. Included in each site evaluation should be the computation of the effects on the surrounding population and property resulting from plant operations and accidents. This evaluation is necessary to determine whether these effects are likely to meet the legal or the generally accepted standards for the optimum or most desirable site.

In the study of the selected site characteristics, a special effort should also be made to identify the required safeguards and control features for the proposed facility. Engineering and biological restraints inherent in the proposed facility systems must be evaluated with the natural environment factors.

To assist in the evaluation of the site, some additional policy questions concerning an estimation of risk of the research to be conducted in the proposed facility must be answered. As a guideline in contrast to the previously mentioned "early planning questions," the following questions are listed in order of decreasing magnitude of risk and decreasing complexity of required precautionary measures. The list applies to diseases of man and animals that might be studied in an infectious disease research laboratory. Because the hazards associated with aerosol dissemination have been demonstrated,[116] and because it is believed that future research will make increasing use of aerosol challenge in the study of respiratory diseases, greater emphasis has been placed upon aerosol dissemination in this list of questions.

(1) Should the facility be suitable for any type of experiment with any microorganism and any animal up to the size of a chimpanzee?
(2) Should the facility be suitable for the preparation of dry powders of infectious agents?
(3) Should it be suitable for the dissemination of pathogenic microbial aerosols of:
  (a) Organisms highly infectious for men, producing a distressing disease for which there is an incompletely protective vaccine and only partially successful specific chemotherapy. The difficulty in treating such syndromes as pneumonic plague causes their aerosolized pathogenic agents to be included at this level of hazard, even though they are not as readily infective as some others.
  (b) Organisms infectious for man, producing disease that is incapacitating but usually not serious when acquired in the laboratory, for which there is an incompletely protective vaccine and no specific chemotherapy. Although the glanders organism is less infective and the disease may be treated with sulfadiazine, it should be included here because of the dangerous clinical syndrome produced.
  (c) Toxins or organisms highly infectious for man, producing disease for which there is either effective vaccination and/or effective specific chemotherapy.
(4) Should the facility accommodate laboratory studies not involving planned dissemination of aerosols? The subclassification would be the same as in (3) above.
(5) Will it be used for the dissemination of dry or fluid aerosols of organisms with comparatively low invasiveness, usually with no vaccine available, often subject to specific chemotherapy therapeutics, but sometimes causing serious pneumonia, such as staphylococcus, streptococcus, and pneumococcus?
(6) Will its use involve laboratory studies not involving dry powders or planned dissemination of aerosols, with organisms of less serious risk because of various mitigating factors present to varying degrees,

such as availability of vaccination, specific treatment, and low infectivity in the laboratory?

(7) Will it be used *only* with microorganisms of minor risk levels such as:
   (a) Nuisance diseases such as Newcastle virus conjunctivitis?
   (b) Organisms seldom causing laboratory infection such as pneumococcus, streptococcus, staphylococcus, meningococcus, vaccinia virus, and diphtheria and tetanus bacilli?

(8) Will it be a facility used only for classroom demonstrations or student work with killed, stained preparations or with attenuated strains?

An analysis of biological waste treatment, decontamination, and disposal facilities should be conducted concurrently with consideration of presently available techniques and local and national applicable codes and ordinances. The local environment may provide protective features to the site by diversifying water channeling or otherwise influencing the movements and diffusion of facility emissions or effluents.

The final site selection should represent the best balance that can be achieved between cost and overall area protection during both normal and abnormal conditions. The results of the preliminary analyses and evaluations are essential information and, together with design criteria concerning separation distances, exclusion areas, safeguards, and containment requirements, should provide data relative to the broad operational controls, limitations, and restrictions.

Biological Contamination Control Requirements

The broad criteria that have been established in preceding sections can now serve as a baseline for a more detailed study and consideration that can be given to possible alternative decisions. A logical approach to this is an understanding of biological or microbiological control concepts and methods which can be expressed in a similar manner to the approach used by a systems analyst in an industrial problem. This approach was provided by Phillips, et al., who in 1965 described five essential stages of biological contamination control.[99] These are explained as follows and shown in Table 2.

TABLE 2. Stages, Approaches, and Techniques of Microbiological Contamination Control

Stage 1 RECOGNIZE AND DEFINE THE PROBLEM

Stage 2 ESTABLISH CONTAMINATION CONTROL CRITERIA

Maximum number of organisms allowed, types of organisms, where located, how detected, and other criteria.

TABLE 2.—Continued

Stage 3	EMPLOY APPROACHES AND TECHNIQUES OF CONTROL				
	Facility design features	Use of containment equipment	Management functions	Use of correct techniques	Use of sterilizing agents, germicides, and other control measures
Stage 4	**MICROBIOLOGICAL TESTING AND SURVEILLANCE**				
	Air sampling	Surface and component sampling	Physical and chemical tests and measurements	Testing of filters, incinerators, sewage, water	Freon leakage testing
Stage 5	**ANALYSIS OF RESULTS AND CERTIFICATION PROCEDURES**				
	Recording results, statistical tests, use tests of items, formal or informal certification.				

Stage 1—Recognizing and Defining the Problem. Problems created by lack of contamination control are often identified in retrospect—some undesired events having already occurred. Contamination control reaches its highest degree of refinement when data are accumulated that allow the problem areas to be predicted and the necessary control measures to be installed *before* losses occur. Once a problem involving microbiological contamination control is recognized it should then be defined as accurately and as completely as possible.

Stage 2—Establishing Contamination Control Criteria. Any attempt to control microbiological contamination lacks significance unless the standards of control that must be achieved are defined. That is to say that there must be a definition, in microbiological terms, of the objective of the control endeavor. Many control operations, for example, require that sterility be achieved and maintained. In infectious disease laboratories the criterion may be to prevent the escape of pathogens. Water treatment systems are concerned with the elimination of pathogens to produce potable water. Food processing plants must render foods microbiologically safe for human consumption. In the hospital operating room certain air-hygiene practices are appropriate to prevent infection of patients.

Contamination control criteria should be established in a manner to facilitate validation of control processes. If sterility is the aim, the criteria should specify what procedures are to be used in testing for sterility, how many replicate tests are needed, when the tests are to be done, etc. The modern trend, resulting from spacecraft sterilization considerations, is to require probability values for sterilization procedures. If sterility is not the objective, the criteria should specify the maximum number and types of microorganisms allowed in an environment, in a solution, on a surface, in a component, etc., and should indicate the test methods to be used. It is most important that this concept be clearly understood. In the food industry and in the manufacture of biologicals, the criteria are generally specified and controlled by certain governmental regulatory agencies. In other areas of microbiological contamination control no widely accepted criteria have been developed. In some areas, such as in planetary quarantine, specific standards for contamination control have been established by the National Academy of Sciences.

Maximum success in future contamination control activities will depend in no small part on continued research in the various fields where control is needed in order to determine appropriate criteria and standardized testing methods. The architect/engineer with a continued interest in microbiological contamination control facilities should follow new developments in applied microbiology for the identification and standardization of control criteria.

Stage 3—Employing the Approaches and Techniques of Control. *Facility Design Features.* Modern design criteria applied in the construction of facilities can do much to control microbial contamination. Understanding and using these criteria is, of course, one of the principal aims of this book. Some of the features that make up the catalog of items that can be suggested for inclusion in new or renovated facilities to control biological contamination are:

(1) Use of ventilated cabinets, chambers, cages, etc., to achieve an absolute or partial barrier to contain microorganisms at their point of use or to exclude them from a specific work area.
(2) Use of laminar air flow devices to exclude microorganisms from a particular environment.
(3) Use of differential air pressures within a facility or sub-facility so that air moves from biologically clean areas toward areas of higher microbial contamination.
(4) Use of appropriately effective microbiological filtration or other treatment for air supplied to and/or exhausted from rooms, cabinets, chambers, cages, etc.

(5) Change rooms, locker rooms, water shower rooms, or air shower rooms for personnel.
(6) Use of germicidal ultraviolet air locks and door barriers to separate areas of unequal microbiological loading or risk.
(7) Treatment of microbiologically contaminated liquid effluents from the facility.
(8) Room arrangement or layout to achieve traffic control within the facility along a clean-contaminated axis.
(9) Use of an effective intercommunication system in the facility to avoid unnecessary movement of personnel from area to area.

For those faced with initiating a design plan of a facility where microbiological contamination control is needed, the problem is one of determining which of the above items are to be used and to what extent. Moreover, it is usually necessary to make these determinations before the planning stage of a new or renovated facility. This is a difficult problem whose dangers are that the facility will provide more contamination control features than are necessary, or fewer features than are necessary, will fail to protect surrounding communities from the contamination, or will be too inflexible in the future to accommodate changes in the contamination control requirements or in the scope of the work.

The best approach to the design of a facility for microbiological contamination control requires consideration of some basic policy decisions before the design is begun. As an example, we have included earlier a comprehensive list of such policy questions relating to the design of laboratory facilities for infectious disease research. It is appropriate to emphasize that the most important major decisions in selecting engineering features for microbiological contamination control should be based on the fact that control should begin at the work surface or that area where the contamination originates or where the item to be protected is located. If this principle is followed, a simplification in design and a savings in facility costs usually results.

Use of Containment Equipment. Experimental evidence and practical experience have shown that handling techniques alone cannot be depended upon for consistent control of microbiological contamination. As the criteria for control become more exacting, aseptic handling techniques fail to provide sufficient containment. Fortunately, however, engineering developments have provided devices and apparatus that provide efficient microbiological and physical separation between environments. The most important type of containment and isolation equipment and the type capable of meeting the most severe control criteria is the gastight absolute barrier enclosure. Ventilated cabinets and animal cages are representative of this type of equipment. A number of recent publications describe the design and use of ventilated cabinets and animal isolation equipment.[14,24,50,99,134] Many different types of *absolute* barrier enclosures have been developed. Cabinets and enclosures also have been designed for the *partial* barrier concept wherein microbiological contamination control is achieved by controlling the direction of the air flow into or out of an open panel on the cabinet. Laminar flow cabinets can perform the same function.

According to the containment requirement, various engineering and performance standards can be established for containment equipment. Such requirements relate to (1) leak tests for the enclosure, (2) ventilation rates, (3) filtration or incineration of air (supply or exhaust), and (4) provisions for decontaminating or sterilizing the interior of the enclosure and the air filters.

An important aspect with regard to containment is the selection of the proper type of equipment in relation to the level of contamination control or the criteria for control. For example, with infectious disease agents used in research laboratories, recommendations have been made that correlate the type of disease agent and the level of risk of the research with the type of protective cabinet needed.[135] (See Table 1.)

In addition to cabinets, chambers, and animal cages, other types of containment equipment are available or can be designed for specific procedures. For example, specially designed containment equipment has been used for procedures such as centrifuging, grinding materials, shaking, blending, and lyophilizing.

Use of Correct Techniques. Even in the presence of adequately designed facilities and good containment equipment, the success of most attempts to control microbiological contamination depends in no small part on the work techniques of the involved personnel. Although no inclusive list of correct techniques would be appropriate for all areas of application of microbiological contamination control, it is possible to discuss some fundamentals that suggest correct techniques and some general types of procedural rules that can be considered.

It is important to emphasize that microbial contamination can exist and yet not be readily detectable in the usual sense; the contamination may be odorless, tasteless, and invisible. Moreover, instantaneous monitoring devices for microorganisms, comparable with devices for detecting radioactive contaminants, are not yet available. Also, it is important to understand the ease with which microorganisms can be made airborne, and to realize their ability to remain viable while airborne in small particulate form and to move from place to place in a facility with air currents. Finally, it is significant that the physical state of a microbiological contaminant is related to the ease or difficulty of containment. It can be demonstrated, for example, that dried, micronized, powdered, or lyophilized microbial preparations are more difficult to contain than contaminants in a wet or fluid state.

In general, "correct techniques" as used in this discussion relate to the movements of people in the working environment insofar as these movements can minimize the spread of contamination through the air or on surfaces. These techniques, moreover, relate principally to the movement of the hands in carrying out work. A meaningful analogy can be made to the techniques of the surgeon and surgical nurse who must at all times be aware of how materials are to be handled aseptically. How materials are handled and the proper sequence of handling are important in controlling contamination. Techniques that involve violent movements, aspiration of fluids, spraying of materials, foaming or bubbling of liquids, and overflow or leakage of materials signal the need for specifying exactly how the technique is to be carried out to achieve minimum spread of microbial contamination.

Use of Sterilizing Agents and Germicides. The following discussion presents a digest of methods for destroying microbial contaminants in order to give the reader some feeling for the problem areas and some useful information for design problems. In spite of the extensive literature and continued research in this field, practical problems continue to arise for anyone employing routine sterilization procedures—problems for which there may be no clear-cut answers. This seeming dilemma is due to the fact that heat, the most reliable means of inactivating microbes, cannot be applied in many situations where the contaminants exist on or in thermolabile materials. It therefore often becomes necessary to resort to less reliable means of sterilization, or to accept something less than sterilization, generally referred to as disinfection or decontamination.

It has long been recognized that chemical decontamination is made difficult by the existence of species differences in susceptibility. In addition, the velocity of the process of sterilization by different chemicals depends to a variable degree on dilution, temperature, presence of organic matter, hydrogen ion concentration, extent of penetration, surface tension, and other environmental factors. The application of germicidal ultraviolet light is likewise limited by its low penetrating power.

Of the numerous physical and chemical means of sterilization or inactivation of microorganisms, those that are most widely applicable can be classified under one of four main headings: (1) heat, (2) vapors and gases, (3) liquid decontaminants, and (4) radiation.

(1) Heat—It is generally accepted that the application of heat, either

dry or moist, is the most effective method of inactivating microorganisms. The exposure temperatures and times required for sterility are generally known and ordinarily can be readily controlled. Whenever possible, heat should be used to sterilize materials. Current textbooks adequately specify conditions for the application of heat for sterilization.[89,103] Recent research on the kinetics of dry heat inactivation of microbial spores has emphasized lower temperatures for longer exposure times for the sterilization of spacecraft and spacecraft components.[18]

(2) Vapors and Gases—A variety of vapors[92,110] and gases possess germicidal properties. Among these are ethylene oxide,[92,110] formaldehyde,[92] propylene oxide,[17] beta-propiolactone,[14,50] and methyl bromide.[19] When these agents are employed in closed systems and under controlled conditions of temperature and humidity, excellent decontamination can result.

Under controlled conditions, ethylene oxide is a highly penetrating and effective sterilizing gas, convenient to use, versatile, noncorrosive, and effective at room temperature. However, the gas is slow in killing microorganisms and usually is mixed with other gases to avoid explosion hazards. Ethylene oxide is widely used to treat items not suitable for heat sterilization. Its use in treating foods is limited because it reacts with and destroys some vitamins and because some of it hydrolizes to ethylene glycol.

The above considerations have accelerated the use of propylene oxide as a sterilizing gas for foods. Some of the foods decontaminated with propylene oxide are cocoa powder, dried vegetables, dry food mixes, dried egg and milk products. Propylene oxide is slower acting than ethylene oxide but it presents fewer toxicity problems.

Formaldehyde and beta-propiolactone are used primarily as decontaminants for room and building interiors. Formaldehyde, the slower acting of the two, has the undesirable property of condensing and polymerizing when sprayed. The polymer, once formed, requires long aeration (sometimes a week or so) for removal. Beta-propiolactone holds great promise as a space decontaminant.[16] In the vapor state, it acts rapidly against bacteria, rickettsiae, and viruses, and has no adverse effect on most materials. It is much faster acting than formaldehyde and does not leave an undesirable residue after spraying. A serious deterrent to the use of this chemical is its toxicity and carcinogenic capabilities under certain conditions.[32,129]

Methyl bromide is about one-tenth as active against microorganisms as is ethylene oxide. This compound has found its greatest use in soil sterilization, especially to eliminate fungi.

Peracetic acid[8] is also bactericidal in the vapor state; however, its primary use is as a liquid decontaminant. The chemical has value in decontaminating enclosures or other areas where a vapor as well as a liquid is required to sterilize the item. Because peracetic acid is corrosive to metals, care must be exercised in the selection of materials to be treated with this chemical.

(3) Liquid Decontaminants—There are many misconceptions concerning the use of liquid decontaminants. This is largely due to a characteristic capacity of such liquids to perform dramatically in the laboratory and to fail in a practical situation. Such failures often occur because too little consideration is given to such factors as temperature, contact time, pH, concentration, and the presence of organic material at the site of application. Small variations in these factors may make large differences in germicidal effectiveness. For this reason, complete reliance should not be placed on liquid decontaminants, even when they are used under highly favorable conditions.

Hundreds of liquid decontaminants or germicides are available under a variety of trade names. Most, however, may be classified as halogens, acids or alkalies, heavy metal salts, quaternary ammonium compounds, phenolic compounds, aldehydic compounds, and other organic preparations. None is equally useful or effective under all conditions.

In the decontamination of large areas or rooms the mechanical removal of microorganisms by washing with water or disinfectant solutions plays an important part. For this reason, surface-active agents are often incorporated into germicidal solutions. The most frequently used liquid disinfectants are chlorine solutions, iodoforms, phenol and related acids, mercuric chloride, formalin, quaternary ammonium compounds, and sodium hydroxide solutions. Solutions of soap must not be overlooked for decontamination purposes.

When decontamination with chemical solutions is required, viruses and rickettsiae present special problems. The evaluation of the virucidal and rickettsiacidal action of chemicals is difficult. Most tests take place under various conditions of time, temperature, pH, and organic material that may be hard to duplicate. Moreover, complete inactivation is difficult to determine because of the methods used for virus detection and assay.

(4) Radiation—Ultraviolet radiation, X-rays, gamma rays, high-energy electrons, protons, alpha particles, and neutrons are examples of forms of ionizing radiation capable of destroying microorganisms. The most common methods presently used for the sterilization of materials (surgical supplies, laboratory supplies, packaged foods, etc.) are: high-energy electrons from a particle accelerator, and gamma radiation from a radioactive source. Although microorganisms vary in their resistance to radiation, a

dosage of around 2.5 megarads usually is sufficient to sterilize surgical materials.[121] Radiation sterilization with gamma rays or high-energy electrons is used mostly with packaged goods.

In certain specific applications, germicidal ultraviolet (UV) radiation at 2537 Å is an effective means of decontaminating air and surfaces. It is sometimes used for the treatment of water and other liquids. Used in air locks or door barriers, UV radiation can isolate areas of differing levels of contamination within a building.[133] It is also useful for reducing extraneous contamination in rooms. Window-type air conditioners used in contamination control areas may be fitted with UV lamps to decontaminate recirculated air. UV radiation has limited penetrating power and thus is most effective on exposed surfaces or in slow-moving air. Proper concentration, contact time, and maintenance are also critical. Phillips and Hanel[97] have adequately described the use of UV for practical decontamination applications.

Management Functions. Management's policies, directives, and other actions are essential to any microbiological contamination control effort. Obviously programs of contamination control are initiated, funded, and supported at the management level. At the time programs are initiated management must assign responsibilities. Persons at various levels in the organization who will be responsible for the outcome of the contamination control efforts must be identified. But beyond this, management at various levels must concern itself with other essential functions. For example, management must be responsible for the proper selection of employees. This refers not only to technical competence and skills but also to the fact that it may be undesirable to employ persons with certain physical conditions or diseases for certain types of work involving microbial contaminants. Management, likewise, should be concerned with providing the necessary training for employees involved in contamination control activities, for formulating work regulations, and for enforcing them.

Stage 4—Microbiological Testing and Surveillance. In the fourth stage of microbiological contamination control, Table 2 depicts five types of procedures for testing and surveillance. In any control endeavor one or more of these techniques is needed to assess whether the techniques employed (Stage 3) achieved microbiological control that meets the criteria established (Stage 2). A useful manual has been published.[39]

Leakage Testing. Testing with chlorofluorohydrocarbons (Freon *) should be used to validate the microbiological tightness of any absolute barrier system. Inability to leak molecules of freon gas is equated with the

* Freon ®—Dupont Chemical Co., Wilmington, Delaware.

inability of microbes to enter or escape from the barrier. For a cabinet or similar enclosure one ounce of freon gas is admitted for each 30 cu ft of space. Compressed air or an inert gas is used to raise the pressure to 6 in. water gauge. There must be no leakage when tested with a G.E. Halogen Leak Detector operating on high sensitivity range.

Testing of Filters, Incinerators, Sewage, and Water. Whenever microbial air filters, air or solid waste incinerators, or sewage or water treatment systems are a part of a contamination control procedure, these systems must be tested to assure adequacy of operation. It is particularly important to test systems prior to their being put into routine operation. In some instances microbiological tests with tracer microorganisms will be appropriate and in other instances temperature measurements and other tests are applicable. Testing must be done in such a manner that a break of sterility or containment is not involved. Decker, et al.[30] have prepared a comprehensive monograph on air filtration and air filters that is recommended for use by anyone concerned with the filtration of airborne microbial particles.

Physical and Chemical Tests and Measurements. According to the nature of the microbiological contamination control endeavor, a number of physical and chemical tests and measurements may be done. In some instances these tests may be critical to the surveillance program and in other instances they may provide presumptive evidence that the control criteria are being met. Whenever wet heat is used for the sterilizing procedure a record of the temperature, pressure, and treatment time should be maintained. With dry heat sterilization the temperature and treatment time must be recorded. When liquid or gaseous decontaminants are used these should be periodically assayed to assure proper chemical concentration and pH.

Surface and Component Sampling. Moistened cotton swabs or Rodac * plates are usually used to detect microbiological surface contamination.[39,53] If possible, the results of surface sampling should be expressed on a quantitative basis (e.g., microorganisms per unit area of surface). Sterile strips frequently are used to quantitate the accumulation of microorganisms on surfaces over periods of time. This has been used for assessing spacecraft contamination levels. Small components may be tested by immersion in an appropriate nutrient fluid or by washing the component in a sterile fluid that is then quantitatively assayed for viable microorganisms. Information on methods for determining the internal sterility of components is incomplete. Obviously, however, these tests must be done in a sterile environment.

* Rodac ®—Becton, Dickinson & Co., Rutherford, N.J.

Special microbial detection and assay tests may be devised for other materials such as oils, greases, and powders.

Air Sampling. Air sampling test and surveillance procedures provide quantitative data on the presence of viable airborne microbes. According to the sampling devices used, assessment can be based on microorganisms per unit volume of air, or on microorganism-containing particles per unit volume of air. Agar settling plates can provide data on viable particulates falling on a unit area of surface per unit of time (e.g., particles per square foot per hour). Other types of samplers can provide estimates of the particle sizes of viable airborne particles. A monograph by Wolf, et al.[136] is an excellent summary of air sampling techniques and devices. The use of selective culture media in air samplers may provide an opportunity to test for specific types of microorganisms. Also, microorganisms obtained from the air during sampling can be subjected to further testing for specific identification.

Stage 5—Analysis of Results and Certification Procedures. It is obvious that the control criteria that are established in the second stage are the guidelines for the analysis of results and certification. Moreover, it follows that corrective actions should be started when a microbiological contamination control process is shown to be out of control or not meeting the minimum standards. In establishing methods for the analysis of results, the following are cardinal considerations:

(1) No biological detection procedure has perfect validity and reliability.
(2) Within certain limits, sterilization and decontamination procedures improve as the challenge microbial load is lowered.
(3) Sampling statistics are a major tool in analyzing the results of microbiological testing.

Consideration of the establishment of the contamination control requirements should be based on the following control functions:

—Prevent Escape —(Prevent Entrance)
—Sterilize —(Decontaminate)
—Eliminate Pathogens
—Maximum Allowed; Where Located
—Detection Methods

The control criteria are valid only if they can be met. Therefore, it is necessary that facility planners performing detailed analysis to develop control processes be familiar with and aware of the following techniques of microbiological contamination control.

Facility Functions. Regardless of the direction of the contamination control effort (product protection and/or personnel protection), there are some common facility considerations that apply.

Enclosures, barriers, or other containment devices that immediately surround the infectious or potentially infectious material are designated as primary barriers. These are the first line of defense (other than the test tubes, flasks, etc.) for preventing escape and possible spread of infectious microorganisms. Examples of primary barriers are conventional ventilated microbiological cabinets, closed ventilated animal cages and laminar downflow biological cabinets (Figure 6).

The secondary barriers are the features of the building that surround the primary barriers. These provide a separation between restricted areas in the building and the outside community and between individual restricted areas within the same building. Examples of secondary barriers are floors, walls, and ceilings, UV air locks and door barriers, personnel change rooms and showers, differential pressures between areas within the building, provisions for filtering or decontaminating potentially contaminated exhaust air, and provisions for treatment of potentially contaminated liquid wastes. These and other secondary barriers provide supplementary microbiological containment, serving mainly to prevent the escape or entrance of biological agents if and when a failure occurs in the primary barriers. Figure 1, page 2, graphically represents the functions of primary and secondary barriers.

Actually, the more effective the primary barriers are, the less need there is for emphasis on secondary barriers. Therefore, during the design phase, it is both important and economically necessary first to determine and select the primary containment devices to be used, thereby reducing the complexity and cost of the secondary barriers.[101]

In developing facility requirements in microbiological contamination control, it is helpful to establish facility zones that relate to the various containment functions. Only one of the functional zones listed in the example below has a containment function. The other zones support the containment zone (or mission area) in several ways: by providing office areas and transitional rooms, by providing a mission support zone, and by providing a zone for the machinery that operates the building. In an infectious disease research laboratory for example, it may be necessary to have two or more zones serving a containment function, especially when large and small animals are used.

Typical functional zones for containment design

—Office and transitional zone

—Containment zone

—Support zone

—Mechanical and maintenance zone

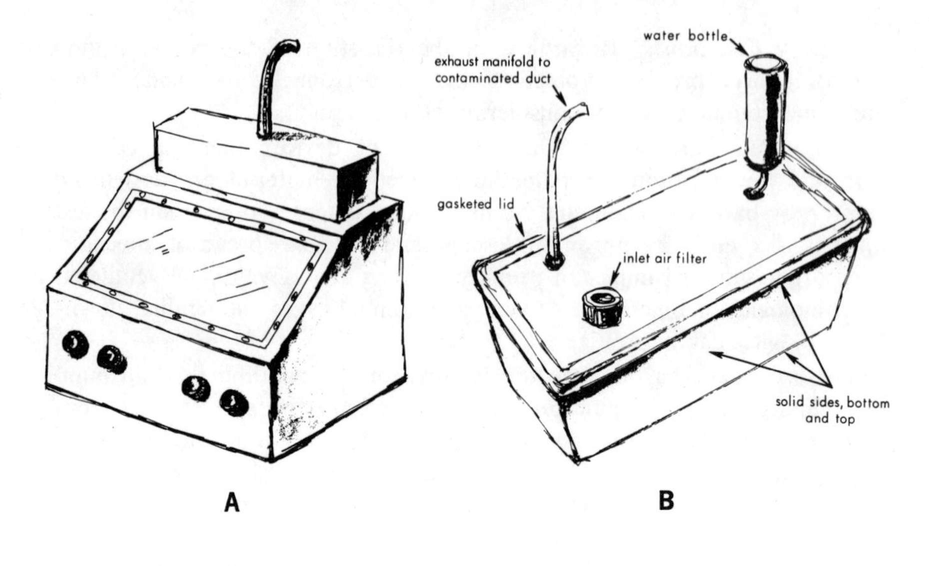

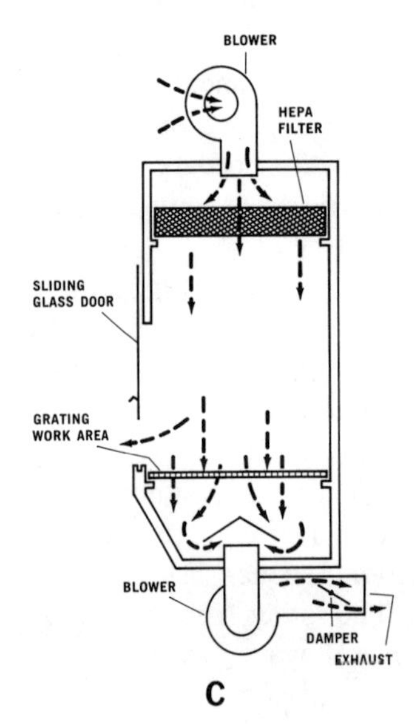

Figure 6. Primary barrier examples: A, ventilated safety cabinet; B, ventilated animal cage; C, laminar downflow biological cabinet.

Knowledge of appropriate construction methods, materials, and techniques is essential for effective control of microbial contamination in a facility. Features to be considered in the planning of new or renovated facilities are summarized on the next page:

Typical features for containment
—Primary/secondary barriers
—Clean rooms (sterile rooms)
—Air balance zones
—Air filtration
—Change and shower rooms
—UV air locks
—Treatment of liquid and gaseous wastes
—Room layout
—Communication systems
—Proper materials for walls, floors, and ceilings
—Constant temperature work or storage areas

For those faced with initiating a design, the process of determining the features that should be used and their limitations will effect savings in both time and effort in later stages.

Use of Equipment. Experimental evidence and practical experience have shown that handling techniques alone cannot be depended upon for consistent control of microbial contamination. As the risk increases (either to product or personnel), the handling techniques alone may fail to provide sufficient containment. Engineering developments have produced devices that can provide efficient microbiological and physical separation between environments. An example of effective type of containment and isolation equipment capable of meeting the most severe control criteria is the gastight, absolute barrier enclosure. Ventilated work cabinets and animal cages are representative of this type of equipment. The design and use of ventilated cabinets and animal isolation equipment has been adequately described.[14,20,50] Facilities have been constructed and described for sterile assembly of spacecraft.[37] Cabinets and enclosures are available for partial barrier application when microbiological contamination control is to be achieved by controlling the direction of the air flow into or out of an open-panel cabinet. Vertical laminar downflow cabinets perform essentially the same function as the partial barrier, inflow cabinets, but in addition in theory increase the isolation within the unit by providing an ultraclean air flow over the work area.

Use of Operational Techniques. In addition to adequate facilities and good containment equipment, control of microbiological contamination will depend in part on the work techniques of personnel. The active participation of a staff microbiologist in the determination of the operational program is essential to a successful facility for microbial contamination control. It must be realized, however, that microbial contamination, unlike radioactive or other types of contamination, can exist and yet be undetected, because it is usually odorless, tasteless, and invisible. Moreover, instantaneous monitoring devices for microorganisms, analogous to devices for detecting radioactive contaminants, are not available.

Following is a list of some items and elements to be considered in establishing barrier techniques:

—Degree of containment
—Purpose and size
—Reliability
—Compatibility with work materials
—Compatibility with decontaminants
—Expected life
—Cost per unit of work area
—Operating costs
—Maintenance costs
—Other

Surveillance Techniques

Coincident to the detailed analysis and procedures, surveillance methods must be identified. Typically they include:

—Air sampling
—Surface sampling
—Component, item, or material sampling
—Physical and chemical tests
—Testing of filters, incinerators, sewage, and water
—Gas tightness testing

These techniques will be key elements in the development of facility requirements.

Certification Procedures

The contamination controls established or designed into a facility will not be valid unless the results can be monitored and certified. In developing certification methods the planners must consider:

(1) Biological detection procedures do not produce perfect validity and reliability, as can a strictly physical monitoring system.
(2) Within certain limits, sterilization and decontamination procedures improve as the microbial load is decreased.
(3) Sampling statistics are a major tool in analyzing the results of microbiological testing.
(4) Four techniques of certification are:
    (a) Recording results

(b) Statistical tests
(c) Item testing
(d) Certification or feedback

TEST REQUIREMENTS

When certain processes or operations are determined to exceed the current "state-of-the-art" of containment techniques, studies or a research/test program may be necessary to fully identify levels of risk and determine operations, processes, and control systems for optimal containment effect. Tests, studies, or research programs which are undertaken must also establish reliability goals for components and systems, special material requirements, and validity of engineering and biological procedures proposed for control and monitoring of containment systems.

Reasons for a test program are:

—Insure or establish reliability
—Prevent accidents—reduce hazards/risks
—Aid in more effective design
—Maintain uniform operational quality or reduce construction costs

The objectives of a test program normally fall into one or all of the following categories:

—Verification of design concepts
—Development of effective operational techniques and procedures
—Determination of performance objectives
—Relation of reliability to end product and mission
—Demonstration of system and subsystem compatibility
—Compliance with control and process manufacturing goals

ACQUISITION, ORGANIZATION, AND PLANS

The early planning activities provide a thoroughly defined concept, basic and compatible criteria, and general schedules for the program. It is also important to define the time-phased technical and administrative responsibilities for management and the control tasks that will be involved in implementing the overall plans. This internal function will be a cost item whether the facility program is industrial, educational, or governmental.

The degree of success in a program is usually directly related to the objectivity and qualifications of the team charged with its prosecution. Dependent on the specified mission organization, the life expectancy of the facility, and the characteristics of its operations, the team may tend to be contractor/consultant oriented, or an adequate competency may be developed within the sponsoring organization. With the latter approach, transitional problems between the design/construction phase and acceptance/operational phase will be minimized. This approach should also facilitate acceptance testing and development of personnel training and orientation programs, and result in optimal on-line operation.

Criteria/Contract Preparation Team

The preparation of work statements for engineering and/or R&D services will require detailed technical, scientific, and contractual inputs. In planning for this activity, it may be advisable to use a consultant. This of course is, in itself, an element that incurs costs and may create scheduling problems. The team approach for preparation of the Facility Program Document is valuable and the use of consultants knowledgeable in biological contamination control can be a significant factor in the correct determination of organizational goals and objectives. Essentially, the planning team should, as a minimum, be comprised of the following professional categories.[83,107]

Scientific. The planning team will require an experienced microbiologist who has sufficient time to participate in all planning and acquisition activities. Qualifications and experience in the planning aspects of new facilities is a rare trait in the scientific community. Therefore, it must be accepted that this important team function (planning for new facilities) will be the responsibility of engineers on the planning team of consultants. The prime function of the microbiologist will be to insure that clarity of purpose and scientific needs are accurately translated into the contamination control criteria and that sufficient factual information is available for preparation of an engineering work statement. It is also normal to anticipate that the projected research may involve not only microbiology, but could also include such diverse fields as metallurgy, veterinary medicine, entomology, parasitology, etc. Therefore, additional or other scientific disciplines may need representation on the planning team.

Engineering. The technical qualifications of the planning team members should encompass the normal disciplines of a design and construction project, augmented by a bioengineer to work closely with the scientific members of the planning team and to minimize communication problems.

Administrative/Contractual. The procurement effort, an important element of planning, is directed toward development of specific and appropriate kinds of contractual coverage. In addition to the contract administration staff activities, technical members of the planning staff must provide guidance in establishing technical and scheduling requirements.

Proposal Evaluation Team

Early determination of staffing requirements for the evaluation of proposals is a necessary planning step. Suitable time will have elapsed in planning to permit source solicitations, preparation of source lists, and source selection consistent with the type of contract established. Often, in complex or large projects, comprehensive preproposal briefings will be required to clarify the technical, scientific, or contractual position with respect to the program. Proper source selection and adequate proposal evaluation requires objective examination of the contractors' response in the following areas.

Scientific. The contractor's proposal should evidence knowledge and awareness of the state of the art of biological contamination control. Often a consultant for a specialized scientific area such as this will be proposed by the contractor, and therefore overall team qualifications are a significant point for evaluation by the scientific member(s) of the staff. Hypotheses, solutions, and established data must be subjectively reviewed and analyzed for program applications and/or relevance. It is clear that the document used in source solicitations must be clear, concise, and must delineate the areas upon which the evaluation will be made. Otherwise, valuable time will be wasted by the evaluation team in the interpretation and analysis of the various proposals submitted.

Engineering. Evaluation of the contractor capabilities should place emphasis on methods of accomplishment, technical competence, performance history, technical staff, and competency in the critical technological areas involved. In every case the review should be done by the staff members most knowledgeable in the particular technical area.

The activities and tasks discussed in the previous paragraphs should provide information for the planning budget and for schedules. The planning budget in particular must include the current facility requirements and provision for facility expansion. The schedule should be based on the mission objectives and the interrelated factors of internal and external activities. Real estate requirements for a project may necessitate acquisition of new property, and the attendant costs and delays must be appreciated in the planning stage to develop realistic budgets and schedules.

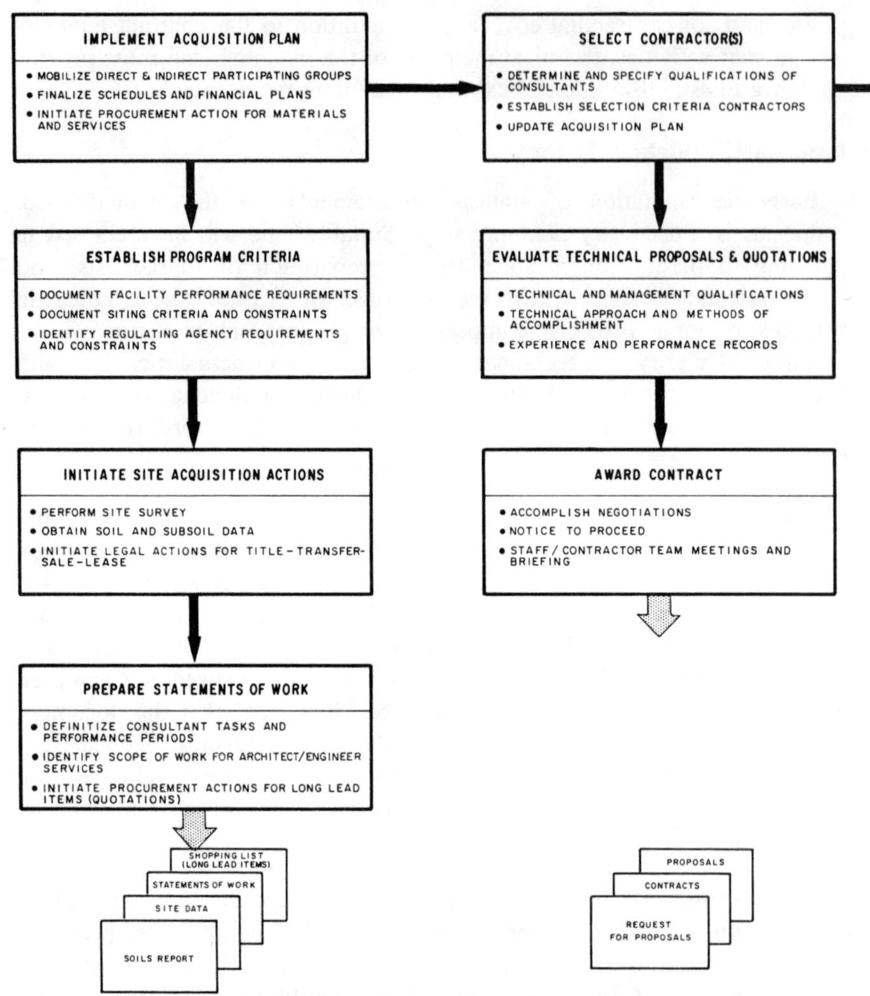

Figure 7. Facility acquisition plan.

Acquisition Plan

Planning functions must result in an accurate "road map" of acquisition activity. Figure 7, a broad and detailed analysis diagram of facility acquisition functions and products, has been prepared to identify significant events. The functions shown are generally sequential, but for given programs many of the activities will be concurrent and must encompass a level of detail consistent with management's interest.

The final step in this planning phase, then, is to assemble, evaluate,

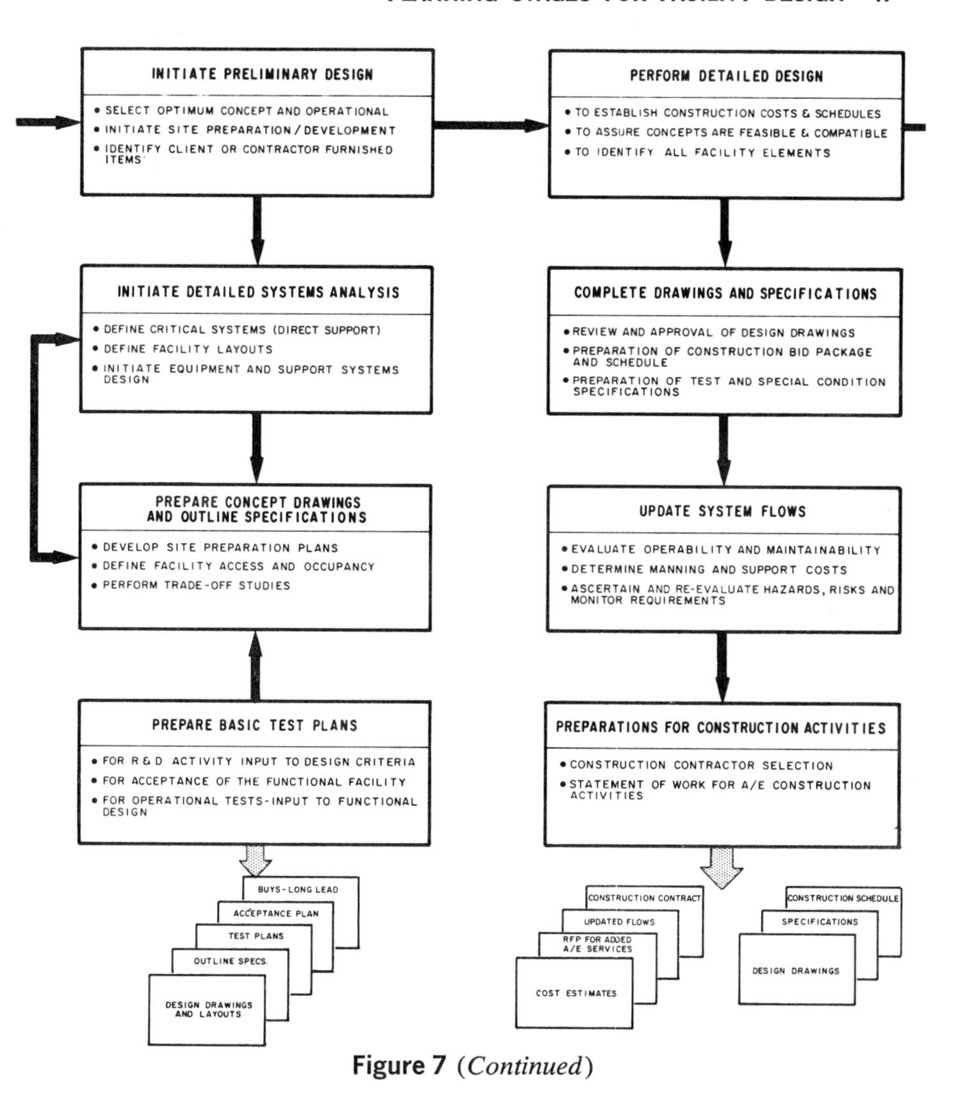

Figure 7 (*Continued*)

and integrate schedules, cost estimates, and organizational plans into a definitive acquisition plan. The plan will be used as a guide for program activities. A network for the interrelated functions also should be an integral part of the plan. It is understood that since all activities will be subjected to change in schedule, cost, or engineering parameters, the "time-phased" road map (PERT or critical path approach) must provide a mechanism for this change to reflect the impact on interrelated activities. Implementation of the plan should begin with mobilization of the man-

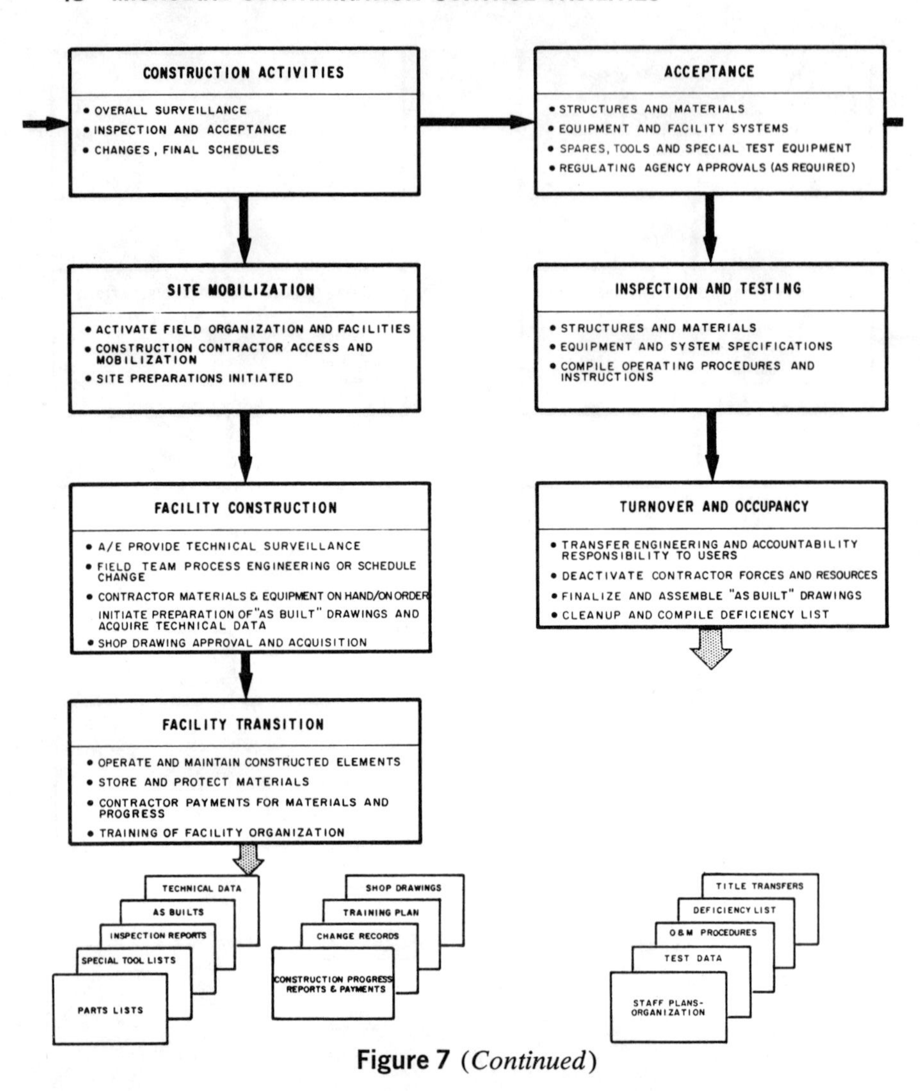

Figure 7 (*Continued*)

agement, administrative, and technical team. The preceding analysis will be a baseline for design activities. Each change, or definable impact on the total plan, must be continuously evaluated against the baseline to assure that mission objectives are not compromised without full consideration of trade-off factors.[72] The contractor's activities provide key interfaces in the plan. His contractual products and schedules should be regarded as significant milestones to be compared with the policies, directives, and other actions that are essential to achieve total microbiological contamination control.

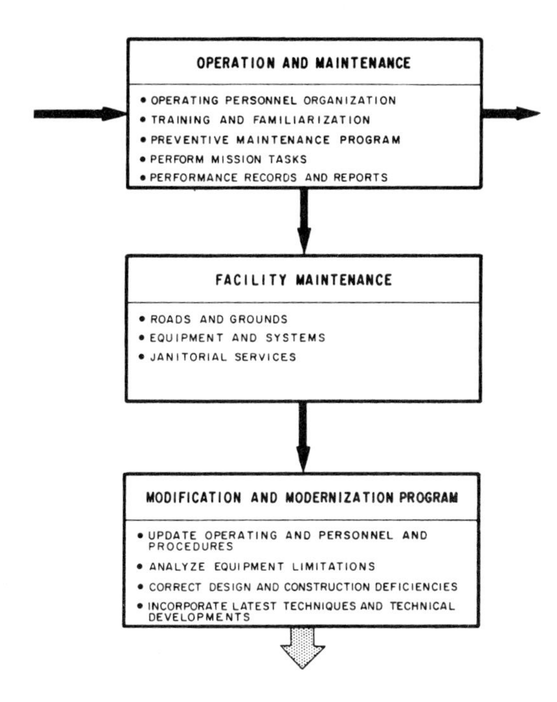

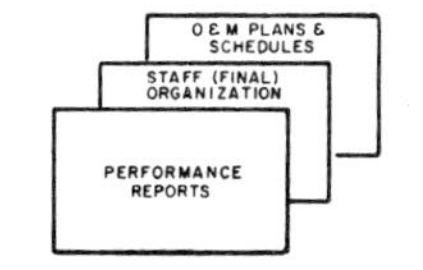

Figure 7 *(Continued)*

SUMMARY

Obviously, the program must be initiated, funded, and supported by the top management level. At the time of initiation, management should assign basic responsibilities and the managers at various levels must concern themselves with the proper selection of employees. This refers not only to technical competence and skills in their field, but also to their ability to analyze and plan for a complex facility program. Management should provide the necessary direction and policies to the team effort on a concurrent basis through periodic review of the planning efforts. An important product of the analysis will be identification and selection of required operational skill levels where human functions are developed from the flow sheets for subsequent program tasks. Development of comprehensive planning charts, such as depicted in Figure 7, should insure maximum coordination and permit identification of the critical stages of facility planning and design.

CHAPTER 3

Selected Architectural Design Criteria

INTRODUCTION

Design criteria for all types of facilities that require microbial contamination control (e.g., research laboratories, biologically clean assembly facilities for the aerospace program, processing facilities in the pharmaceutical industry, and isolation facilities in hospitals) must, of necessity, be broad in scope. Only general and broad outlines covering principles of biological isolation, control, containment, and safety can be discussed. As previously mentioned, a number of recent documents give detailed information on specific facilities and on specific systems.[20,31,37,40,47,57,83,98,101,102,106,108,112,113,118,124,135] There is, however, a considerable body of general principles of building design, equipment design, and containment techniques that are applicable to work with biological materials. These principles and guidelines should be considered by those responsible for planning and designing facilities that require microbial contamination control.

Design criteria for microbiological facilities can be a valuable aid to management, technical personnel, and architect/engineers in planning for microbial contamination control. These criteria will remind personnel planning such facilities that solutions to various containment problems must be considered and will provide an outline for successful solutions to these problems. Previously (pages 15–21), several questions were raised concerning required management policies that may be helpful. If properly used, these criteria may prevent or minimize overdesign of the facility to meet a specific containment situation. Overdesign will almost always result in higher initial costs and higher operating and maintenance costs, and may result in a facility that is restrictive and burdensome to the technical and operating personnel. For example, in the design of a laboratory building for microbiological research, scientific personnel on the planning team frequently request incineration (or burning) of all air from the building. In the average building, such an incineration system would, of course, be enormously expensive to purchase, install, and operate. Furthermore, it is usually not needed because adequate removal efficiencies can be obtained with microbial filters.[27,28,30,56]

In the specific case of infectious disease laboratories, techniques, procedures, and equipment likely to create microbial aerosols hazardous to research personnel should, in fact must, be isolated in ventilated cabinets, hoods, or other containment equipment. Occasionally, but infrequently, air exhausted from these special containment enclosures will be best treated by incineration. But the exhaust air from the general laboratory should almost never require incineration because personnel will be working in this laboratory and will be exposed to the same air that would be passed through the incinerator. Conversely, in a product-protection situation such as drug or sterile disposables production area, it would be unwise to provide sterile air to the entire facility if the product is maintained in a closed system with its own filtered air supply.

One of the latest engineering developments for biological safety and control is the use of streamlined, unidirectional air of minimum turbulence in partial barrier cabinets, rooms, or systems for increased environmental control and safety. Laminar flow is the term generally used to describe such equipment. In opposition to turbulent flow, laminar flow is defined as streamlined flow in a viscous fluid near a solid boundary. In laminar air flow equipment, the entire body of air within a confined area moves with a uniform velocity of 90 ± 20 ft per min along parallel lines in a manner so as to produce a minimum of turbulent air patterns.

This laminar air flow concept is currently being used in hospitals, both in surgical theaters[40] and in patient rooms;[130] also in the microbiological assay of space hardware[78,102] and during the preparation of tissue cultures.[25] Additionally, laminar air flow has been used for contamination control during sterility testing procedures in the production, filling, and packaging of drugs and pharmaceuticals,[11] and during the rearing or use of specific-pathogen-free or defined-flora animals or for protecting normal animals under test from zoonotic disease.

ARCHITECTURAL AND STRUCTURAL CRITERIA

General Principles

Many developments and much new knowledge have been gained in the field of architectural and structural design of scientific facilities. Support given to modern science enables design engineers to utilize up-to-date design principles developed by intensive studies of laboratory practices and needs. The scientist by and large is no longer required to move into old, outmoded buildings that were not originally designed for research work.

Biomedical buildings are seldom units in themselves, but are usually parts of a building complex, and the building is usually part of an installation, research center, or campus. Therefore, one important consideration for the architect is to locate the building in relation to the campus or installation and to coordinate the exterior appearance of the proposed facility with a well considered master plan of the installation or center.

Flexibility is another important design objective for the architect. The rapid development of new equipment and practices makes it desirable to permit future changes in room or suite arrangements quickly, conveniently, and economically. For example, the flexibility of a laboratory will be increased if it is designed on a modular basis. However, its unit cost may not be reduced if maximum flexibility is provided.

The following information from a recent publication by Norman [86] explains some concepts of space saving and cost reductions achieved with modular design concepts.

Cost reductions, which are the measure of efficiency in constructing any facility, are achieved by repeating, in various combinations, a two- and sometimes three-dimensional unit of space, termed a "module." The space needs of facilities can be initially studied to help determine the size of a planning module.

There is a basic difference between an architectural module for construction and a special facility planning module. The former relates to a unit of space having a special length and width or radius, and sometimes height, from which can be generated such architectural and structural elements of a building as spacing of colums, bays, and windows, and location of interior walls. Dimensions of architectural modules may be derived from any one or more controlling factors.

A facility planning module however, is specific in that it is defined as the smallest repetitive unit of space within which is found all architectural, mechanical, electrical, and other functional requisites for the operation of a complete and environmentally self-sufficient research entity. This definition implies, for instance, that an independent research laboratory can occupy more space than a laboratory planning module, but never less. What features the planning module includes will depend on the types of facilities needed and the extent to which policy, in the light of financial considerations, may relax the above definition.

Figure 8 illustrates the essential space elements of the most popular combination of rectangular planning modules for biomedical laboratories. This has a width that has been modified to include a double work depth and a single point of access for the two backed-up work depths. Figure 8 shows how structural and other interferences can restrain the use of the

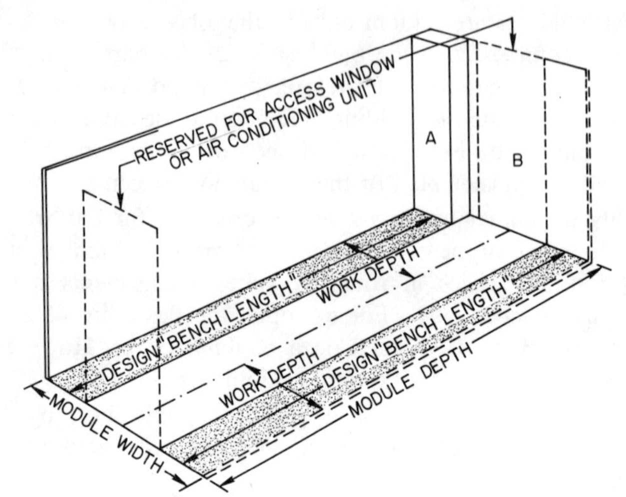

Figure 8. Space elements of a facility planning module.

full periphery of the module. Element A depicts an encroaching column or enclosed (furred) pipe chase that limits the design bench length along the wall. Similarly, element B shows how other repetitive architectural or mechanical features may interfere with the placement of counters or floor-mounted scientific apparatus along a wall.

Obviously, a planning module of such pure definition is applicable only in those facilities where there are no columns to interfere with locations of corridor doors and where there is a maximum flexibility in bringing utilties up through the floor or down from the ceiling. The most popular interpretation or definition of the facility planning module, the one shown in Figure 8, is: a unit of space that has a width consisting of a double work depth, potential access from the corridor, and a full complement of required utilities and environmental features. This definition of the facility planning module more readily permits it to coincide with the architectural module.

Module width. Elements that determine width include:

(1) Thickness of dividing walls or partitions.
(2) Method of distributing utilities within module, including utility curb, if required.
(3) Depth of counter front to rear.
(4) Width of aisle.
(5) Number of laboratory occupants.

Normally, the decision that exerts the greatest leverage in fixing the width of a module is the selection of aisle width. Aisle width depends

TABLE 3. Aisle Widths, Based on Anthropometric Data for Average and Large Men [a]

Activities in Aisle Opposite Each Other	*Dimensions* [b]	
	Average Men	*Large Men*
1 working (sitting or standing) and 1 passing.	43.2 in.—adequate for normal-sized people only.	48.1 in.—adequate for all subjects.
1 working and 1 rising from sitting to standing position (18 inches allowed for chair).	57.4 in.—not quite adequate.	63.2 in.—60 in. considered adequate.
1 working and 1 bending (36 in. allowed for average bending).	43.2 in.—adequate for normal-sized people only.	48.1 in.—adequate for all subjects.
2 working and 1 passing.	61.9 in.—60 in. inadequate.	69.2 in.—66 in. considered adequate.

[a]Based on table in reference 87.
[b]3 in. allowed between people while stationary and on either side of those passing.

primarily on the proposed number of occupants for the module and is influenced by considerations of safety and convenience.

The Nuffield Foundation report utilized data based on body measurements that, when combined with certain assumptions, were developed into a rational method for sizing the aisle.[87] These data are shown in Table 3.

The combined thickness of a partition and the depths of opposite counters are less important than the width of the aisle in fixing module widths. Partition thicknesses normally range from about 2 to 6 in. Front-to-rear depths of counters alone or with utility curbs normally range from 24 to 30 in. Hence the combined maximum variation in depths of opposite counters, plus a partition, is about 20 in. By comparison, the widths of aisles normally vary over a range of 30 in. (from 4 ft to 6 ft 6 in.).

Selection of a partition thickness of 3 in. and a counter plus utility curb depth of 30 in. allowed development of comparative module widths based on the Nuffield table of aisle widths. These are shown in Table 4.

Module Shape and Length. Normally, in rectangular-shaped facilities, ratios of net to gross areas can be increased by increasing the ratios of module lengths to widths. This is due largely to the fact that corridors, stairwells, and other public and maintenance areas remain at relatively

TABLE 4. Module Widths Derived from Body Measurements, Assuming Partition [a] and Counter Dimensions [b]

Activities in Aisle Opposite Each Other	*Module Width*	
	Average Men	*Large Men*
1. 1 working and 1 passing	8′10″	9′3″
2. 1 working and 1 rising from sitting to standing position (18 in. allowed for chair)	10′	10′6″
3. 2 working and 1 passing	10′5″	11′

[a] 3-in. thickness.
[b] 30 in. front to rear.

fixed amounts in relation to increases in useful space, which is the area within the planning modules.

Additionally, the usual single module design relegates the bulk of interior furnishings, such as hoods, sinks and drainboards, counters, and floor-mounted scientific apparatus, to the module walls that are perpendicular to the corridor. Any increase in the length of the module compared with its width represents a net gain in space along the side wall. This space is generally the most suitable for connecting apparatus to utilities. While there are practical limits to the depth of a module, the two reasons cited are responsible for the rectangular shape of most modules, with their long axes set perpendicular to the corridor.

The length of a module can be determined by one or more of the following procedures or restrictions:

(1) Adopting the modular lengths of structures occupied by similar research organizations.
(2) Repeating an existing modular pattern of a structure contiguous to one being planned.
(3) Identification of major building function (e.g., a dominant research discipline or hospital bedrooms.)
(4) Maximum number of staff assigned to a module, including the individual investigator.

Two basic ways to obtain flexibility for facility design purposes are: (1) selecting optimal dimensions for the module, and (2) planning architectural, electrical, mechanical, and other design features that permit division of the module into smaller units of space.

Insets A, B, and C in Figure 9 illustrate three concepts of flexibility in facility space assignments made possible through a combination of proper

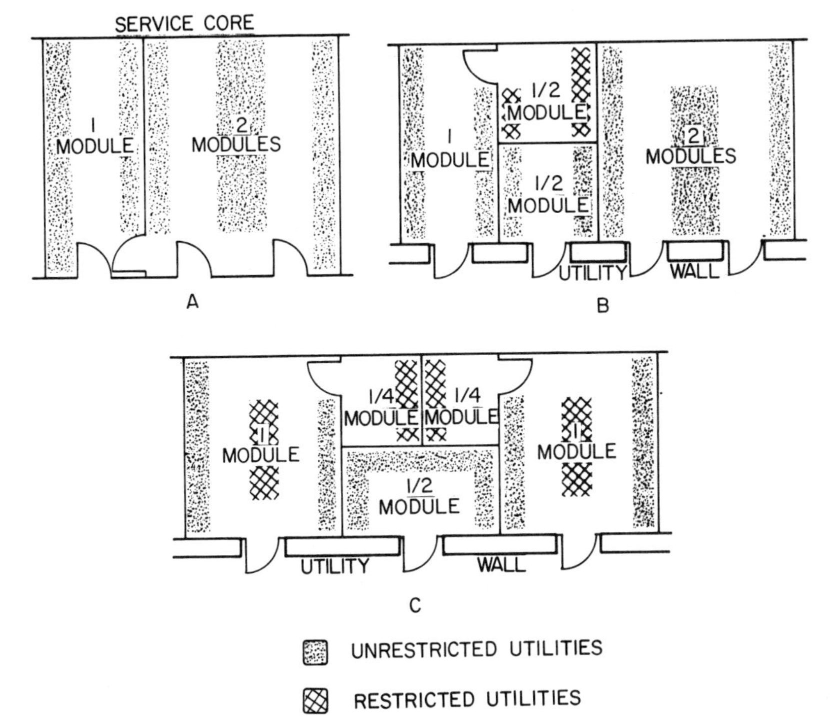

Figure 9. Laboratory planning modules.

selection of module size and location of mechanical utilities. Almost all facility designs could fall into one of the three generalized plans.

Utilities are furnished to the modules in inset A from the rear wall, which may be an outside wall or utility service core. This arrangement is frequently used where modules are back to back in windowless areas. One disadvantage of this scheme is that internal communication is possible only adjacent to the corridor.

The general design of inset B, Figure 9, permits maximum independence of the half-module adjoining the corridor. This half-module may even qualify as a planning module if all utilities are available from service walls lining the corridor. This scheme permits assignment of space for independent laboratories on the basis of $\frac{1}{2}$, 1, and $1\frac{1}{2}$ modules, or larger.

The general design of inset C presupposes that utilities are furnished from points along the corridor. While it requires a wider module, it has the added feature of potentially providing usable facility space in increments of $\frac{1}{2}$, 1, $1\frac{1}{4}$, and $1\frac{1}{2}$ modules.

Whether a fairly generous depth of module or shallower depths will more nearly fit projected needs and the financial resources available is a basic point to be decided. This requires a determination of the space needs and frequency of occurrence of laboratories for a dominant research discipline or for some other selected controlling factor.

Building Design

If a facility is to be less than four stories high, and if it is not to be located in a seismic area, wall-bearing construction should be considered. Such buildings have no columns or beams to interfere with casework or pipes and in most instances will provide greater flexibility at a lower cost than buildings designed within a structural frame. The construction costs of a facility will be less if the architectural and framing plans are straightforward, if materials are of standard size and limited in types, and if the cutting of masonry is kept to a minimum.

The size of the facility should be optimum for the number of occupants, and should include allowance for future growth based on past growth rates or known trends. For example, in microbiological and viral research laboratories, about 300 sq ft per research worker should be allowed; and in planning clinical and diagnostic laboratories, it may be best to specify total size in terms of linear feet of wall space for workbench and equipment. Because wall space is the most important space in a diagnostic laboratory, planning problems should be evaluated in terms of their effect upon this space for benches and free-standing equipment. This approach has been termed the "bench concept." [86]

It is evident that each type of facility will have its own special problems and therefore may have significantly different design criteria for biological contamination control. The facility planner must be responsible for consulting competent recognized authorities to determine the special design criteria and to become familiar with known and/or projected future trends for the particular discipline in question. It is also necessary to become familiar with existing and projected state and federal regulations pertaining to the particular type of construction.

The initial design can be developed to allow for growth of the building in several ways. A building can be laid out so that a mirror image can be attached along one wall, preferably the back. In this manner, the building could be doubled in size with the same room arrangements. In some complexes it may be desirable to design a small, compact basic building. Later, as growth conditions dictate, buildings of the same design could be constructed and connected to the first structure with open or covered walkways, if warranted. These buildings can be arranged in various geometric patterns. This unitized type of structure is often used by the military in

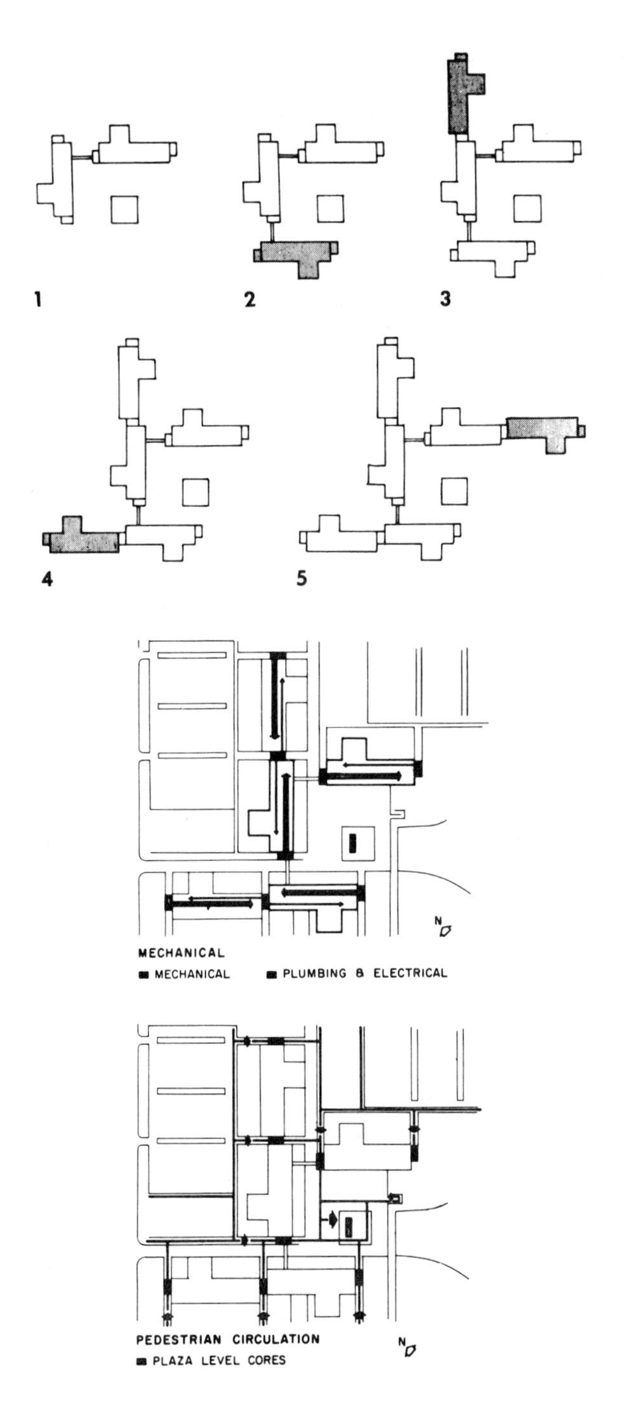

Figure 10. Laboratory complex showing growth from two to six buildings.

constructing temporary station hospitals. Universities and industrial laboratories also find this design economical and flexible. An illustration of growth from a two to a six microbiological laboratory complex with circulation patterns and service pathways is shown in Figure 10.

Floor Plans

In general, multi-story buildings are more economical in terms of initial cost than single-story buildings. Cost of land and available space often will be the determining factors. In addition, single-story buildings can become so spread out that it will take longer to go horizontally from point A to point B than it would to go vertically from A to B by stairs or elevator. In microbiological facilities, traffic patterns should be arranged so that personnel proceed from clean areas to areas of increasing contamination or hazard.

Although there are many different design arrangements for animal breeding and production facilities, two arrangements have proven particularly effective in increasing production rates and controlling microbial contamination. The "clean and refuse" corridor system, illustrated in Figure 11, utilizes dead-end service corridors that greatly reduce the transfer on contaminants. The "clean corridor" provides circulation for operations such as food and bedding delivery and for removal of animals to research areas. The "refuse corridor" provides a means of controlling contamination arising from operations such as removal of dirty cages, dead animals, or used bedding.

The second system, the "separate building concept," illustrated in Figure 12, utilizes many small buildings, each containing one or more

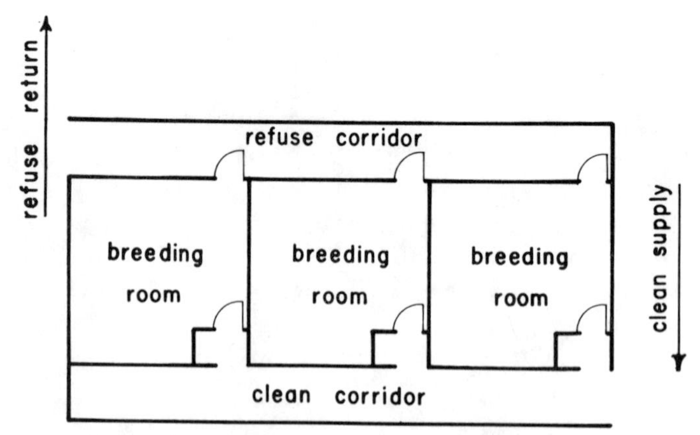

Figure 11. Clean and refuse corridor layout.

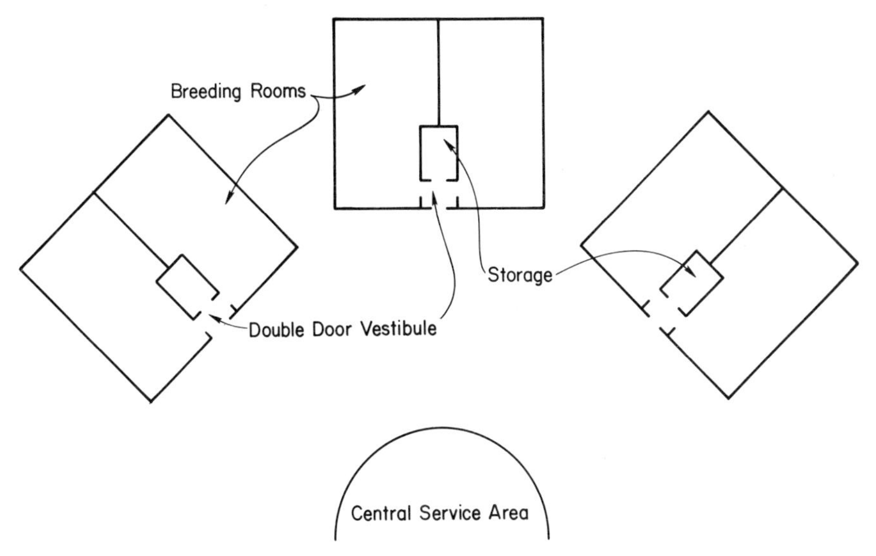

Figure 12. Separate building concept of breeding facilities.

breeding rooms, mechanical equipment, and storage space. The small buildings are located around a central service building that serves as a receiving building for supplies, and office area, and provides a cage and rack washing area. In this concept, the open space between each building, coupled with segregation of animal care personnel by building, provides adequate biological contamination control.[108]

An additional approach that has not been adequately evaluated but that seems promising provides contamination control at the cage level. This is accomplished by using individually ventilated cages,[23] or filter bonnets on each cage top.[66] Manipulations of defined-flora animals may need to be conducted in special clean areas or possibly laminar flow cabinets. If this concept proves feasible, it may be possible to provide sufficient biological contamination control, with minimal facilities expenditure, to maintain virus-defined or germ-free animals.

Some standard design features have been developed for facilities for research with highly infectious microorganisms. The principal feature is isolation of the work areas from the clean areas by interposition of change rooms. The change rooms (for men and women) should not be of the same size but should be in a ratio of 60/40 or 70/30. This flexibility will make it possible for the predominant sex to utilize the larger change room. An illustrative layout of a microbiological research laboratory utilizing this zone arrangement is shown in Figure 13. Offices are provided in the labora-

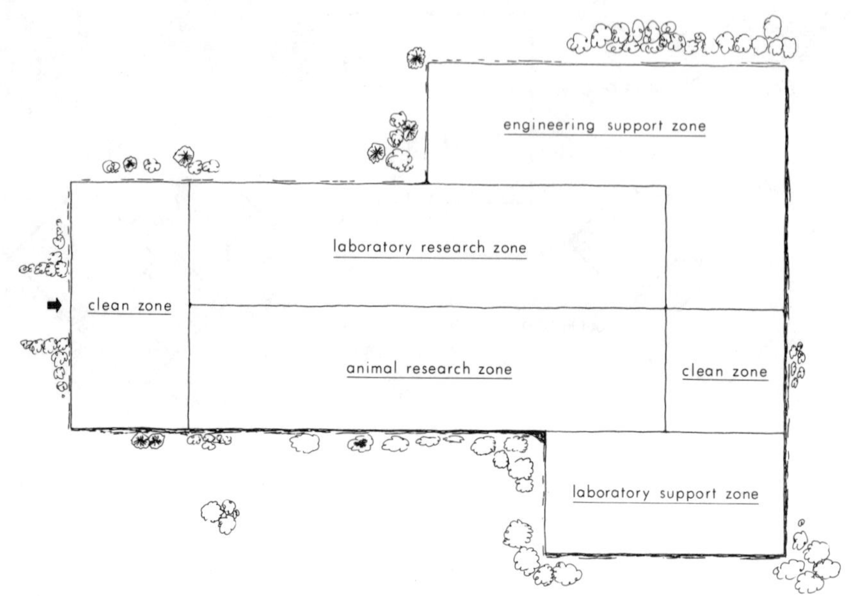

Figure 13. Illustrative building layout.

tory research zone on the contaminated side of the change room because movement from the contaminated side to the clean side requires a shower and a change of clothes. One of the most important items in this type of isolation facility is good communications between the two areas. Speaking diaphragms and viewing windows are required, as well as public address and telephone systems for easy communication. Depending upon the size of the building, service areas such as cage washing and cage storage can be located inside or outside the contaminated area. In a small building these services would be located within the contaminated area because a separate work crew could not be maintained in the clean area specifically for these services. However, in a large building these services, including glassware and animal cage washing, might be placed on the clean side of the building or they might even be located in a separate services building.

Facilities for production of drugs, and sterile disposable medical items, and for those activities in the aerospace program that require assembly of spacecraft with reduced biological contamination, can be designed in a manner similar to the infectious disease laboratory, except that such facilities need to prevent *ingress* of microorganisms rather than *escape*. In these facilities, traffic patterns will be from contaminated areas toward the clean areas, and differential air pressures will be from clean areas to contaminated areas. The contamination control goals in these facilities are similar to those faced by the microbiologist who works with germfree animals or the virologist using clean tissue culture techniques.

Furniture and Equipment Layout

It has been recognized that bioengineering studies are needed for facilities involving biological contamination control and that a specific problem is the lack of sufficient reference material in a suitable form.[86] Biological safety problems have, however, been solved to a large extent with developments of various ventilated cabinets, hood systems, and other containment equipment. However, many of these safety devices and much of the available casework are awkward to use and produce inefficient work arrangements.[34] Figure 14 shows body measurements relevant to bench spacing and some recommended heights and clearances for laboratory furniture. Several studies on the efficient utilization of animal holding facilities have been carried out,[59,108] although these are primarily studies of normal or noninfected animals. Also, glassware and cage washing and media preparation activities have been studied, and recommended layouts, procedures, and techniques are recorded.[38,84]

The problem of ventilated suits for biological contamination control facilities has had very little organized research effort.[12] The new concept of laminar air flow in hoods, systems, and rooms has had extensive engineering input,[41] and is now being evaluated and use-tested in a bioengineering approach.[65] An interesting study of the human factors restrictions of closed ventilated glove enclosures was recently completed.[34] However, additional effort is still needed.

Laboratory furniture (casework, chairs, desks, stools, etc.) should have an acid-, alkali-, and solvent-resistant finish. In general, steel equipment with an appropriate factory-applied finish that is not adversely affected by steam has proven superior to wood equipment in laboratories requiring frequent steam/disinfectant decontamination. Work surfaces (tabletops) can be fabricated of natural, quarried stone, "Alberene," of an impregnated composition asbestos, "chemstone," of carbonized birch, or of a wood surface finished with a variety of coatings from epoxy resins or laminated plastics to stainless steel.[125] A watertight seal to floor and wall surfaces should be employed for workbenches, cabinets, and other pieces of installed equipment, unless the items are easily removable to permit rapid, easy decontamination. It is recommended that all laboratory storage cabinets, lockers, wall-hung cabinets, and similar furniture that does not extend to the ceiling, have tops that slope forward to prevent buildup of dust and to facilitate drainage after washdown with decontaminating solutions.

Body measurements relevant to bench spacing.

Activity	*Average*	*To allow for 97 per cent of population*
(1) Working position	18.7 in.	21.1 in.
(2) Walking between benches	18.5 in.	21.0 in.
(3) Bending (derived from arm length correlated with trunk length)	48.8 in.	52.8 in.

Activities in gangway carried out opposite each other	*Dimensions*			
	Average men		*Large men*	
(1) 1 working and 1 passing (sitting or standing)	43.2 in.	Adequate for normal sized people only	48.1 in.	Adequate for all subjects
(2) 1 working and 1 getting up from sitting to standing position (18 in. allowed for chair)	57.4 in.	Not quite adequate	63.2 in.	60 in. considered adequate
(3) 1 working and 1 bending (allowing 36 in. as average bending, *not* 48 in.)	as in (1) above		as in (1) above	
(4) 2 working and 1 passing	61.9 in.	60 in. not enough	69.2 in.	66 in. considered adequate

Figure 14. Body measurements, bench spacing, and recommended heights.

Special Areas and Rooms

Animal Facilities. In February 1967, the U. S. Department of Agriculture published regulations for implementation of P. L. 89–544, commonly known as the Laboratory Animal Welfare Act.[68] These rules and regulations currently apply primarily to animal dealers and to research facilities receiving funds under grants, or contracts from the federal government for the purpose of carrying out research, tests, or experiments with animals. The U.S.D.A. regulations require compliance with the standards for the humane handling, care, treatment, and transportation of animals and specify some specific restrictions concerning cage sizes and environmental control. This law is the first of its kind in the U. S. and it is probably too soon to determine what the ultimate effect upon the design of animal facilities will be. The law does not require as stringent standards for facility

Recommended heights and clearances for laboratory furniture.

Type of bench	Bench height	Seat height	Minimum vertical clearance from ground to underside of bench
Sitting only	2 ft. 4 in. (28 in.)	1 ft. 5 in. (17 in.)	2 ft. 2 in. (26 in.)
Standing and sitting	2 ft. 10 in. (34 in.)	2 ft. 1 in. (25 in.)	2 ft. 8 in. (32 in.)
Standing and sitting	3 ft. 0 in. (36 in.)	2 ft. 3 in. (27 in.)	2 ft. 10 in. (34 in.)

	Minimum horizontal clearances under bench		
Type of bench	At bench level	At ground level	Minimum knee-hole width
Sitting only	1 ft. 6 in. (18 in.)	2 ft. 0 in. (24 in.)	1 ft. 11 in. (23 in.)
Standing and sitting	1 ft. 6 in. (18 in.)	2 ft. 0 in. (24 in.)	1 ft. 11 in. (23 in.)
Standing and sitting	1 ft. 6 in. (18 in.)	2 ft. 0 in. (24 in.)	1 ft. 11 in. (23 in.)

Figure 14 (*Continued*)

design as are recommended by other government agencies.[61] Future amendments, however, may increase the scope and requirements beyond those currently stipulated.

Probably the most valuable medical research tool, with the exception of research equipment and instrumentation, is the laboratory animal. In order to obtain accurate and consistent experimental results, scientists must have at their disposal animals of a uniformly high quality. Animals form the baseline of a great many experiments. If this baseline varies, subsequent findings will reflect this deviation. Appropriate biological contamination control, proper housekeeping, disease control and administration, and adequate animal housing facilities are required to provide this uniform supply of high-quality research animals.

Scientists, engineers, and others concerned with the welfare, health, and pedigree of the laboratory animal are constantly upgrading the quality of the average research animal through improved husbandry practices and physical facilities. Therefore, the following definitions of the three basic types of research animals currently available may, through progress, become outdated. Those animals raised in facilities with no special environmental controls, and routine housekeeping practices, are classified as *conventional animals.* Because minimal environmental and biological contamination control practices are employed, these animals may have clinical or subclinical signs of disease. *Specific-pathogen-free* (or virus-defined animals) are generally of a higher quality than conventional animals and are raised and maintained in facilities requiring more complex contamination control techniques. Animals identified as specific-pathogen-free (SPF) are maintained without certain known organisms, and they are periodically tested and certified or guaranteed to be free of these specific pathogens or organisms. *Germfree animals* are those animals known to be free of life for which biological tests are available or are employed. Another category of animal, known as *axenic* (or defined-flora) is basically an SPF animal of extremely high quality. These animals are produced by obtaining germfree animals and then recontaminating them with one or more desirable organisms.

The distinctive qualities of these types of animals are primarily maintained by biological contamination control techniques, i.e. the use of increasingly complex physical barriers as one changes from the use of conventional to germfree animals. In the design of animal facilities for research laboratories, it must be recognized that emphasis on different species may change from year to year and that there has been and probably will continue to be more extensive use of the pathogen-free, germfree, axenic, and defined-flora animals. Therefore, the facilities should, if possible, have built-in flexibility to allow a maximum of different uses for any one room or area. The paragraphs that follow deal with the design of facilities for the production and breeding of laboratory rodents, for the quarantine of dogs, cats, and primates, and for the holding of rodents and larger animals under experimentation. Particular emphasis is placed upon the biological contamination control techniques and design criteria.

Production and Breeding Facilities. Animals bred in standard production facilities are mice, guinea pigs, rats, hamsters, and rabbits. Because these animals can be maintained at relatively similar environmental conditions and housed in cages on movable racks of a similar size, the design criteria for housing for these species are similar. In many cases, an organization will alternate the use of its animal rooms to accommodate all of these species. If a research institution produces its own animals, rather than purchasing them from a commercial source, either the production facilities or the conditioning spaces should be close to the research complex.

This close proximity will greatly reduce the difficulties of transportation of the animals, a process which can create an unwanted environmental stress. The production area may be separated from other buildings and confined to one portion of the property so that unauthorized personnel may be excluded from the buildings, and noises and odors from the animals can be isolated.

The different design arrangements for animal breeding and production facilities, the "clean and refuse" corridor system, the "separate building" concept, and "cage level" contamination control have been previously mentioned. These are several distinct areas required in an animal production facility. Of these, the animal breeding room is the most important. The size of this room may vary between 300 and 700 sq ft. A typical room should contain a sink, a workbench, some storage space, and electric outlets. If the breeding rooms are oversized, epizootics could destroy the entire room, which might be a large portion of the total colony. If the room size is maintained with the stated range, optimum care and control are provided. A small entrance vestibule for hand washing is helpful for preventing cross-contamination. Provision of an area for animal caretakers to change clothes and to shower before entering the breeding rooms is an important means of reducing the entrance of microbial contamination.

The type of animal caging system can affect room size and layout.[88] If cages for the different animal species can be standardized so that one basic size cage rack is used, this rack unit can then be considered the basic facility planning module discussed earlier (pages 53–58). In this approach, design of the building and selection of ancillary equipment such as rack and cage washers can be simplified. The possibility of using fixed racks with removable cages should be investigated. Floor cleaning is facilitated if fixed racks are attached to the wall and do not touch the floor. If cages with grid bottoms and waste pans that can be flushed are used, labor and handling costs usually can be reduced. This system has been used successfully for rabbits, rats, and monkeys, as illustrated in Figure 15. The figure shows a screened vestibule required for primates, a garbage disposal under the floor to prevent clogging, and an automatic watering device.

The need for storage space should not be underestimated. Animal feed, bedding, cleaning equipment, racks, and cages can occupy considerable space. The cage washing area should be conveniently located with respect to the breeding rooms. Space should be provided for soiled and clean racks and cages. A loading dock leading directly into an enclosed short-term storage room is needed for receiving shipments of feed, cages, and other supplies. (Tables 5 and 6 contain information concerning average amounts of bedding and food required on a basis of pounds per animal per month.)

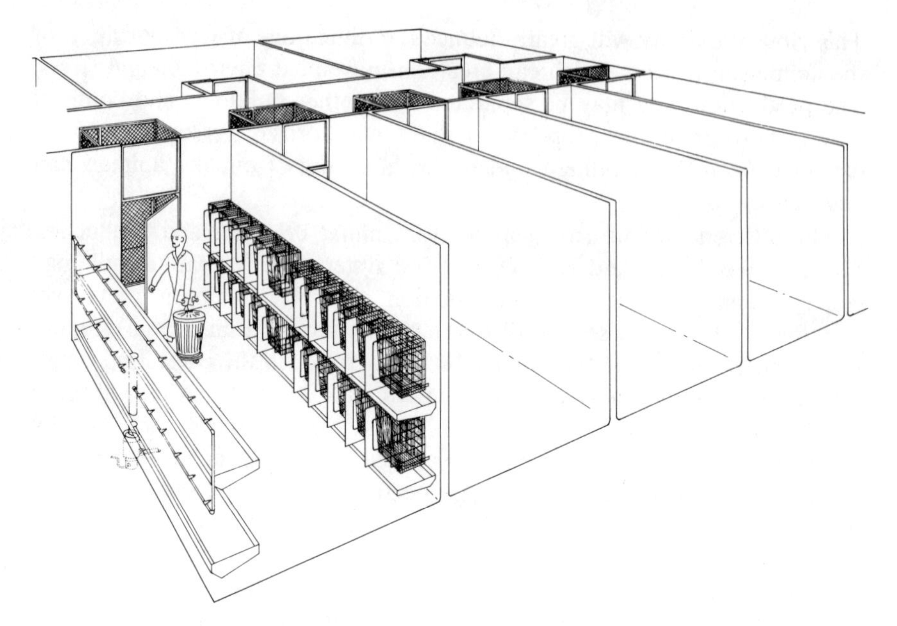

Figure 15. Automated features of animal rooms.

TABLE 5[a]

Species	*Animals/ Month (Average)*	*Lbs Bedding/ Month (Average)*		*Lbs Animal/ Month (Average)*	
Mouse	109,000	9,435 Kg	20,000	86.2 g	.19
Rat	38,300	11,748 Kg	25,900	304 g	.67
Guinea pig	10,350	20,956 Kg	46,200	2.02 Kg	4.46
Rabbit	3,400	7,167 Kg	15,800	2.11 Kg	4.65
Hamster	4,850	1,814 Kg	4,000	374 g	.824
Dog	700	—	none	—	none
Cat	160	454 Kg	1,000	2.84 Kg	6.25
Monkey	1,100	1,814 Kg	4,000	1.65 Kg	3.63

[a]From production records, Animal Production and Animal Hospital Sections, Laboratory Aids Branch, DRS, NIH.

TABLE 6 [a]

Species	*Daily (g)*	*avdp (oz)*	*Monthly (lbs)*	
Mouse	5	0.175	0.4	181 g
Rat	15	0.53	1	453.6 g
Guinea pig	30	10.5	2	907.2 g
Rabbit	150	52.5	10	4.53 Kg
Hamster	10	0.35	0.7	318 g
Dog	500	175.0	33	14.9 Kg
Cat	150	52.5	10	4.53 Kg
Monkey	300	105.0	21	9.5 Kg

[a]Ralston Purina Co., St. Louis, "Purina Laboratory Animal." 1962(—), p. 32.

While this loading dock can be used for shipments of animals, it is more desirable to provide a separate dock for this purpose. Addition of a pathology laboratory and autopsy facility will aid in the early detection of disease in the colony. Provision should be made for employee shower, locker, and lunch rooms. See Figure 16 for a diagram of the area relationship.

The production of specific-pathogen-free animals will require virtually the same types of rooms, areas, etc., but the biological contamination control criteria will be more rigid. Several authors have recently described such facilities.[42,48,131] The production, rearing, and use of germfree animals

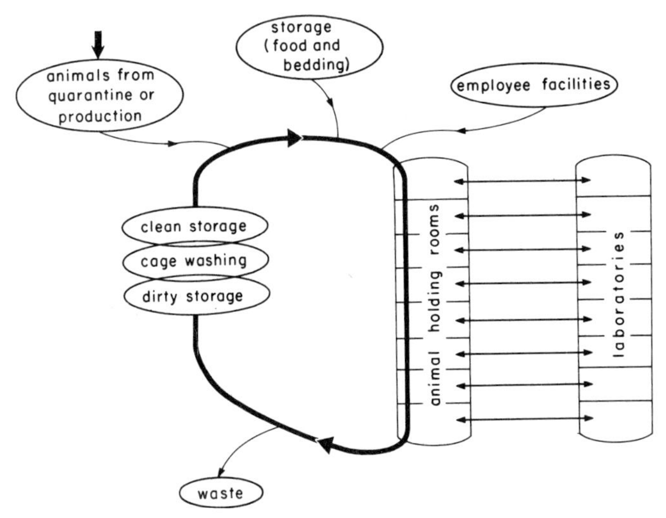

Figure 16. Area relationships.

are currently conducted entirely behind physical barriers, either rigid metal or plastic, or flexible plastic units. Recent articles concerning the design and operation of such facilities are available.[8,52,59,71,106,123,124]

Quarantine Facilities. The arrangement shown in Figure 16 could also be used for a quarantine facility for examining, treating, and conditioning animals prior to their use.

If no breeding stock is maintained, an area should be reserved for inspecting and conditioning the animals received at an institution. This area should preferably be separated physically and operationally from other portions of the animal colony.

Quarantine rooms for laboratory rodents are not different from rooms in experimental areas that are to be described later. Small animals such as mice, rats, and other laboratory rodents, if not produced by the organization, should be purchased from dealers who are well known and the quality of their animals predictable.[127] Therefore, this section will describe facilities for only dogs, cats, and monkeys. These animals require a long period of quarantine; and, because they are often transported many miles and may be in poor condition of health when received, medical and surgical treatment may be needed.

In quarantine facilities, animals are moved from the receiving dock, to conditioning rooms, to the holding areas, prior to transfer to research space. Materials are moved from the receiving dock to storage. Separate covered platforms for shipping and receiving animals should be provided. The receiving platform should be located adjacent to the quarantine area. There should also be adequate storage space adjacent to the receiving dock. A separate preparation room for animals to be shipped is also desirable.

A room should be provided for examining and dipping or spraying the animals as they enter the quarantine building. This room should be completely washable with a well sloped floor and floor drain. The room should contain cages for animals to shake and dry in after being dipped or sprayed.

Dogs that are on long-term holding should be provided with an exercise area. If space is available, an indoor-outdoor run arrangement is a very convenient method of housing the dogs. The Laboratory Animal Welfare Act stipulates cage and run sizes on the basis of length of the animal and then translates this into square feet. Therefore, generalization on the ratio of weight/square feet is no longer permissible. It is undersirable to house more than four dogs in a runway. However, when dogs are placed in a run together, they should be of similar size and of the same sex. Gang cages in which many dogs are housed together are more difficult to keep clean and fighting often becomes more of a problem. Figure 17 shows the relationship of the interior run to exterior runs.

Figure 17. Interior and exterior dog runs.

In a quarantine building housing all types of dogs, a solid partition should be provided on the lower 1.21 m to 1.52 m (four to five feet) of the runs so that waste cannot be transferred from one pen to the next. This also prevents much of the barking caused by visual contact between animals. A grate-covered waste gutter should be provided at the outside end of the outdoor runs. The runs may be modified for holding cats by putting a wire covering over the top of the outside run. Hot water radiant heat coils in the indoor run will dry the floors quickly after washing and help keep the animals healthy. The climate in which the facility is located will determine whether facilities for snow removal from the outside runways should be provided.

A quarantine period, which will vary among institutions, is required for primates. During this time outdoor runs are not required; in fact, individual cages are preferable. These may be either movable cages with solid sides, top and grid bottom, or the stainless steel or galvanized wire type that are placed over a stainless steel waste trough. This latter method of caging has become very popular in the short time it has been in use. Automatic watering devices and automatic flushing devices for the waste pan reduce caretaker cleaning time considerably.

An important aspect of the quarantine area is a multipurpose operating room. This should be available for minor surgery and for required treatment of animals purchased from pounds or other sources. Provisions should be made for employee locker, shower, and lunch rooms.

Animal Experimental and Holding Areas. Rooms of this type may be required to house almost any type of animal under experimentation in close proximity to research laboratories. The individual details and arrangement of the room are dependent upon the type of research and the species of the animal housed. Three basic arrangements are used: animal rooms adjacent to the laboratory; animal rooms in one area on each floor; and/or a separate animal building adjacent to the laboratory building connected with a walkway. The first system is usually not economical because the rooms are dispersed over a large area and facilities for storage and mechanical equipment must be unnecessarily duplicated. In addition, problems of biological contamination control are increased by such dispersion. However, such a concept may be required if the scientist must personally inspect his animals frequently, as in an infectious disease research facility, where the animals may be an integral part of the research program and maintained behind a biological barrier system.

Thought should be given to the use of separate elevators and corridors for human and animal traffic if animals are housed in a laboratory building. In the design of this type of facility, a number of precautions should be observed. It is important, for example, that floors be waterproofed to prevent liquids from leaking into laboratories or other operations on lower floors. Cracks in walls and ceilings should be sealed to prevent harborage by insects and disease-carrying agents.

If these rooms or areas are to be used for small primates, several specialized problems or concepts must be recognized. A wire mesh vestibule within each room will assist in preventing animals accidentally leaving their cages from escaping into other parts of the facility. In addition, a suspended wire mesh ceiling may prove desirable in rooms that have high ceilings, suspended light fixtures, or exposed ductwork, which provide ideal hiding places for escaped primates. Careful consideration must also be given to cage design to prevent the animals from opening the doors to their cages.

In addition to provisions for animal cage and rack washing, provision should be made for the removal of soiled bedding and the receiving and storage of clean bedding, water, and feed. Very often it is feasible to transport soiled animal cages and cage racks to a central washing area where the bedding is removed and racks, cages, and water bottles are washed and then reassembled. With some pathogens, however, this process may be a great hazard to caretaker personnel and a mechanism for spread of disease through the faciilty.

The second arrangement of animal holding space, with all animal rooms at one location on each floor, eliminates the duplication of facilities inherent in the first system, although it does present the same construction problems. A possible way of eliminating some problems is to locate all animal quarters, cage and rack washing, and storage rooms on the ground floor. This method can eliminate vertical traffic of animals and animal equipment and thereby minimize the possibility of cross-contamination.

The most economical and practical arrangement is the third, which separates the animals entirely into an independent building, provided that it is satisfactory to the investigators involved. When all animal facilities are grouped in one building, duplication is at a minimum, cross-contamination problems are reduced (since the animal buildings and the laboratory are separated by a natural barrier of air), and service and storage areas can be centrally located.

The inclusion of an operating suite and supporting areas in the separate animal building, such as are required for corrective and experimental surgery, will facilitate the work of the scientists. The operating suite should include a preparation room, operating room, recovery area, examination and treatment room, X-ray and autopsy room, pharmacy, and pathology laboratory. Provisions also should be made for employee locker, shower, and lunch rooms. Figures 18 through 22 illustrate the different areas of an animal operating suite.

When animals have been experimentally infected with a disease that is transmissible to man, special handling techniques and equipment are needed. Safe handling of infected animals can be accomplished by isolating the animal in a ventilated cage or cabinet. Air is drawn into the cage or cabinet through a filter and is withdrawn in such manner as to create a reduced air pressure within the enclosure. (See Figure 23.) As an alternate method of safety when handling infected animals, workers can be protected with respirators, ventilated head hoods, or ventilated suits. In experimental animal rooms holding large numbers of animals, it may be impractical, cumbersome, and time-consuming to care for the animals in individually ventilated cages. In such situations, the animals possibly can be held in open-topped cages, employing an ultraviolet light screen to prevent cross-infection,[96] with the worker wearing individual safety devices to protect him from the organisms in use. This is a major decision that should be made only after data on animal cross-infection have been reviewed.[62,94,116] If only respiratory protection is required for adequate safety, the worker may wear a device such as is shown in Figure 24. This respirator is equipped with a plastic drape to protect the worker's skin against exposure to ultraviolet rays. If more protection is needed, an air-supplied head hood can be worn as shown in Figure 25. A maximum degree of biological protection is afforded by the air-supplied ventilated suit shown in Figure 26. This suit may be decontaminated as the worker leaves the experimental animal room by use of a 2% peracetic acid shower solution in a stainless steel or plastic-coated shower stall.

Animal rooms used for the study of infectious diseases should be located in a remote part of the building. It is often desirable to have a contaminated corridor running along the back side of the animal rooms. This arrangement may be visualized by referring to Figure 27. Personnel may enter the contaminated corridor through a narrow change room that connects it with

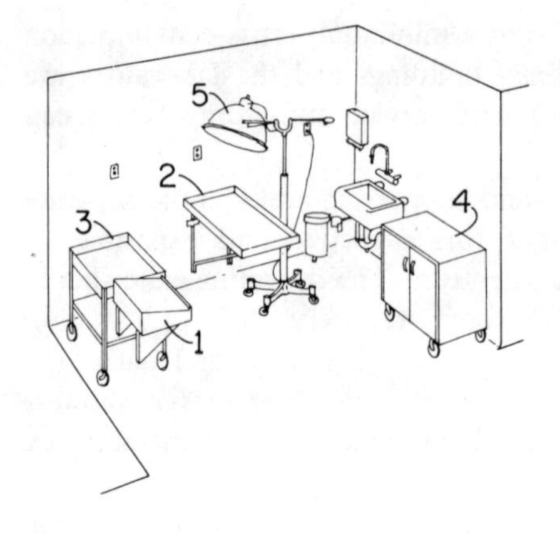

Figure 18. Operating suite. Examination alcove: 1, wall-hung writing shelf; 2, wall-hung examination table—removable; 3, soiled instrument cart; 4, clean instrument storage; 5, movable examination light.

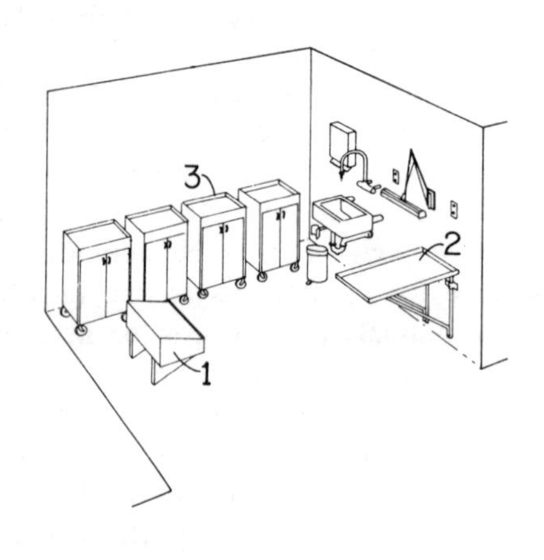

Figure 19. Operating suite. Treatment alcove: 1, wall-hung writing shelf; 2, wall-hung examination table—removable; 3, treatment carts.

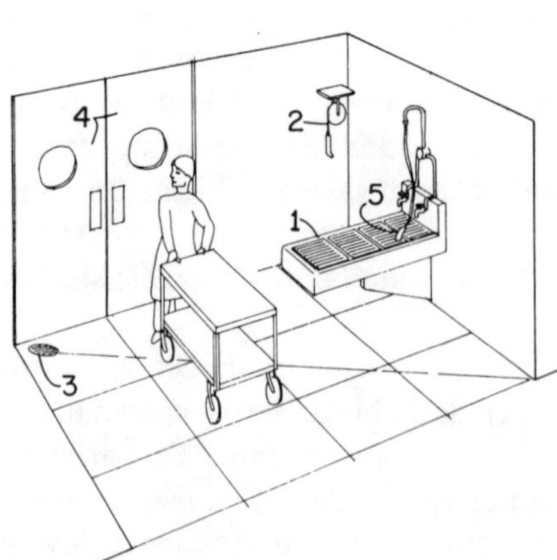

Figure 20. Operating suite. Preparation alcove: 1, preparation table; 2, hair cutter; 3, floor drain; 4, doors to operating room; 5, spray hose with mixing valve.

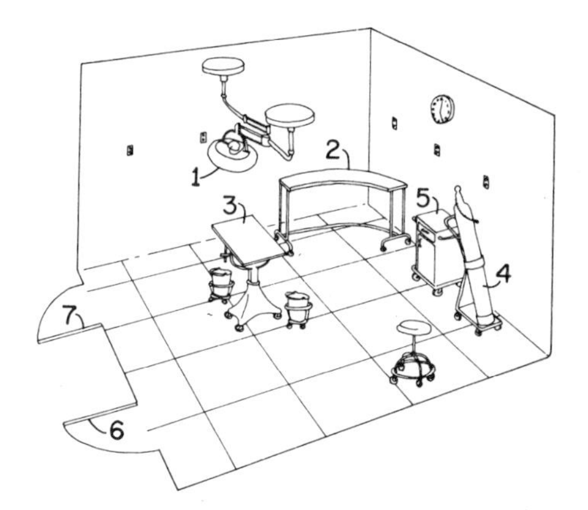

Figure 21. Operating suite. Small operating room: 1, operating lights; 2, instrument table; 3, operating table; 4, oxygen tank; 5, storage cabinet with casters; 6, door to sub-sterile room; 7, door from scrub-up.

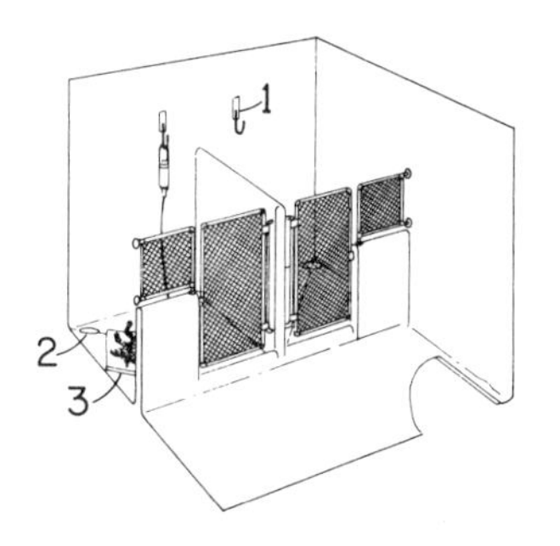

Figure 22. Operating suite. Recovery room: 1, hook for blood plasma; 2, floor drain; 3, pallet.

Figure 23. Ventilated animal cages. (*U.S. Army photograph.*)

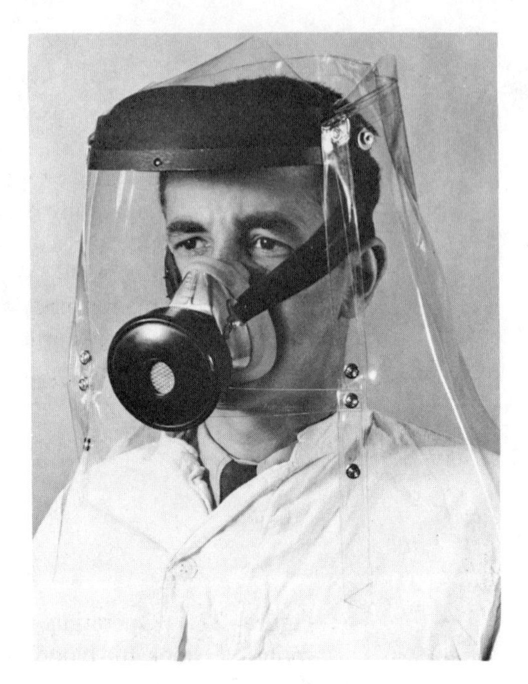

Figure 24. Human respirator with plastic shield. (*U.S. Army photograph.*)

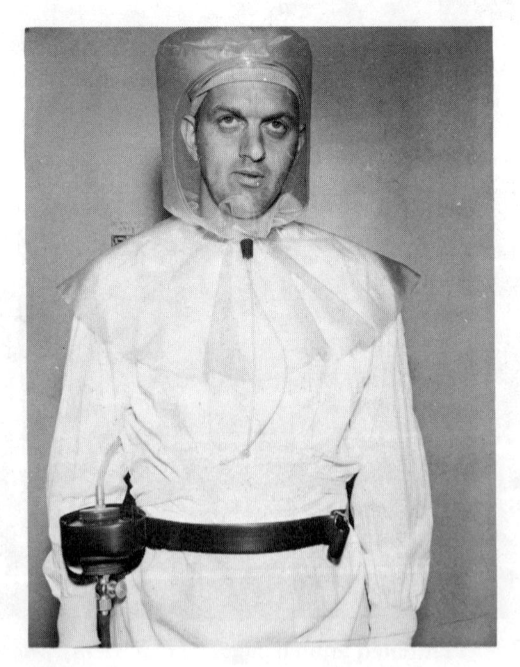

Figure 25. Air-supplied personnel hood. (*U.S. Army photograph.*)

Figure 26. Ventilated suit. (*U.S. Army photograph.*)

the main corridor. A feed storage room can be located between the main corridor and the contaminated corridor. Ventilated animal autopsy cabinets can be extended from the contaminated corridor wall to the main corridor wall. There can be a pass-through air lock extending from the contaminated hallway into the autopsy cabinets, and a double-door autoclave that extends from inside the autopsy cabinet through the wall into the main corridor. At the end of the contaminated corridor, one or more double-door autoclaves can be installed through the wall so that cages and other equipment can be sterilized as they are moved from the contaminated corridor to the cage washing and storage room. In this arrangement, (Figure 27) the main corridor is maintained as a noncontaminated corridor similar to all other corridors in the building.

Physical Design Factors for Animal Facilities. After the information concerning the function of the animal facilities and the preliminary design concepts of the major areas has been assembled, the following factors

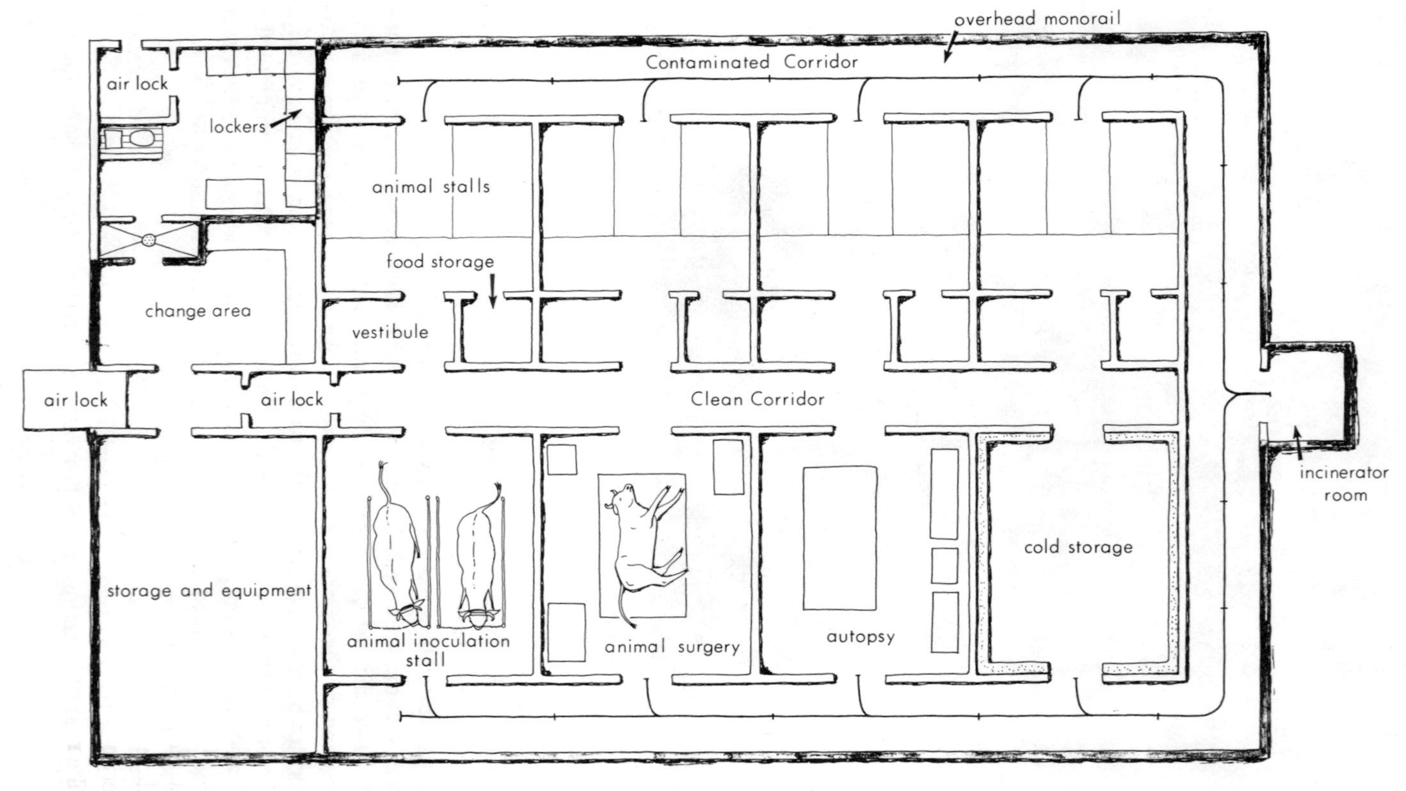

Figure 27. Animal room layout for infectious disease study.

should be evaluated for each area: architectural needs; lighting and electrical needs; mechanical needs; and the need for automation.

The interior surfaces of the animal rooms must be easily cleanable and must have many of the same properties as a laboratory. The floor and wall surfaces, however, must be even more durable than the laboratory finishes because stronger detergents are used to remove the dirt that collects in animal rooms. The amount of cage and equipment traffic is also heavier than the personnel traffic.

The floor is one of the most important considerations for animal room sanitation. Both construction materials and maintenance requirements are considerations. Bacteria that attach themselves to dust particles and settle to the floor will multiply if suitable nutrients and moisture are present.

Perhaps one of the most widely used flooring materials in animal rooms is concrete. It is a very satisfactory material if it is troweled smooth and treated with a surface hardener and sealer. Construction specifications should be very exacting with regard to surface designation and workmanship. Terrazzo is somewhat more durable than concrete. However, it is also more costly to install and repair. Nevertheless, in corridors where traffic is very heavy, consideration should be given to the installation of terrazzo. Quarry tile, used in many installations, is resistant to the organic acids and salts in animal wastes. Two disadvantages are the difficulty in cleaning because of the joints, and the noise that is created as wheeled vehicles are moved through the corridors. Some of the plastic-type surfaces that are troweled, sprayed, or rolled over concrete in a thin layer resist wear and cracking somewhat better than plain concrete. The resiliency of some of these materials also makes them significantly quieter than the harder-surface finishes. However, a definite problem is that of obtaining good supervision to insure proper application of these special coatings. Membrane waterproofing should be used whenever animal facilities are located on a supported slab over occupied areas.

Wall surfaces of animal rooms should be smooth, hard, impact-resistant, free from joints, and resistant to urine, cleaning compounds, and decontaminating solutions. Many wall materials provide some of these qualities, but compromises usually have to be made either because of cost or since a material with all the desirable features is unobtainable. Glazed tile has been used for many years in areas where a high degree of cleanability is required. It is durable, but unfortunately it will fracture if subjected to impact from vehicles. Moreover, the finished joints are rough and difficult to clean. The use of an epoxy grout, however, may render the joints impervious.

Film materials that are applied with adhesives and overlapped to form a tight joint will usually give an impervious surface. These materials are flexible, so the danger of cracking or joint separation is reduced. Many of the new synthetic coatings, designed to be applied over concrete or cinder blocks, produce a durable surface at a moderate cost. However, care must be exercised in the selection of the particular brand since there is considerable variation in their properties. The concrete block and mortar should be free of any unstable compounds that might cause a separation of the synthetic coating. Cement plaster on a cinder or concrete block wall will result in a fairly durable and smooth wall. Any type of hung or false ceiling provides a location for objectionable insects and infectious organisms. This can be overcome by using a flat concrete ceiling finished with an appropriate sealer and/or paint. A finishing operation may be required to produce a smooth surface. With this type of ceiling, air ducts may be surface-mounted and caulked at the ceiling to reduce dirt-collecting surfaces. Alternatively, the ducts may be eliminated from the room and placed in the corridor ceiling for easy access without disturbing the animal population.

Doors in animal areas should be at least 3 ft 6 in. (1.08 m) wide by 7 ft (2.13 m) high to accommodate animal cage racks and service carts. Push plates and recessed door pulls should be used rather than doorknobs. In the planning of corridor widths, consideration should be given to animal cage and rack dimensions. An investigation should be made of the feasibility of using sliding doors. The use of sliding doors, suspended from a track and mounted either on the corridor side or the room side of the wall, allows elimination of the area required for the swing of normal doors and reduces damage to door hardware from moving carts.

Doors in corridors and in supply aisles may be equipped with photoelectric devices for automatic operation. Gaskets should be installed at the bottoms of outside doors to prevent entry of wild rodents or pests. Because such gaskets will render doors airtight, exhaust louvers in corridor doors or walls will facilitate accurate balancing of air pressures. This is particularly important for isolating animal room odors and possible cross-contaminations. Doors located in areas that are washed down frequently should have some means of protecting the doors against rust. A stainless steel kickplate on both sides and the bottom is one recommended solution.

A curb suitable for protecting the wall from the impact of wheeled vehicles might be installed in animal rooms and corridors of buildings specifically designed for animal housing. A curb 6 in. (15.2 cm) high and projecting 4 in. (9.1 cm) from the wall with a sloped top to eliminate dust-catching surfaces should be sufficient. If desired, a large radius cove may be used instead of curbing. A curbing or cage bumper should always be used with plaster walls.

A major consideration in the design of animal areas is the control of noise. This is especially important when the animal facilities are located in the same building as laboratories or offices. Noises in an animal facility generally originate from two sources: the animals themselves, and transportation of the cages and racks from one area to another. The main problem is encountered with the barking of dogs or chattering of primates. Where animal noise would create a problem an attempt should be made to isolate the animals in rooms. Some laboratories practice debarking as a temporary measure; however, this is not a permanent solution to noise control. Keeping doors closed reduces noise transmission. The use of rubber tires and a smooth resilient surface in heavy traffic areas will reduce noise. The architect and engineer should realize, however, that probably the most effective means of handling noise transmission is to separate the animal quarters physically from any area that would be disturbed by noises.

Lighting and Electrical Factors. Adequate electrical outlets should be provided, preferably of the grounded, waterproof type. All rooms should be provided with normal 110-V service, and 220-V service should be available in areas that use heavy-duty equipment.

Light fixtures of the fluorescent type are preferred and these should be watertight to permit hosing. They should be either surface-mounted and caulked or recessed to eliminate a horizontal, dust-catching surface. The design range for light intensity should be 60 to 100 ft-c. Use of multiple switching in each room allows a change in intensity for different functions. Cycling devices for the lighting systems in the animal breeding rooms is very useful. All light switches should be located outside of animal rooms for convenient access and protection from moisture. Sunlight is not required for small-animal breeding rooms and is undesirable from the standpoint of the added heat load. Direct sunlight on small animals confined in cages may cause temperatures to rise above a safe level.

Mechanical Factors. The following plumbing requirements should be thoroughly investigated by planners of the animal facility. Hot and cold water will be required for all facilities; however, many institutions will have additional needs for demineralized water, steam for sterilization, compressed air, gas and vacuum piping, and special sewerage provisions.

Use of floor drains in small-animal rooms will depend on whether it is desired to hose down the rooms regularly, to sweep and damp-mop them with a disinfectant, or to install or use a portable wet vacuum system. It is cheaper to install floor drains during initial construction than to add them at a later date if the room's function is changed. Rooms with floor drains are more flexible in that they can be used for small and large animals or for procedures where larger volumes of water may be required for cleaning.

If drains are installed but not needed immediately, a removable gasketed cover may be placed over the perforated cover. If drains are likely to become clogged with waste feed and bedding, a flushing floor drain or a garbage disposal unit in connection with the drain can be considered. Even when floor drains are omitted, materials that are easily washed should be provided, since damp-mopping of floors and walls with hot water and detergent or wet vacuuming is necessary for adequate germicidal protection. In addition, decontamination of the room with chemical vapors may be periodically required. In rooms where fixed cage racks are used in conjunction with flush-type waste pans, a drain line must be used to dispose of waste water. Use of a closed drain system, coupled with membrane waterproofing, will greatly reduce the possibility of leakage between floors when animals are housed over other research functions. Adequate provision should be made for clean-outs in a closed drain system.

Floor drains are usually essential in rooms in which cages are hosed down. This type of cage sanitation is very efficient and dependable, and the animal house must be designed to accommodate such operations. All solid wastes may be flushed from the cage into an open gutter and then into a drain. It is recommended that the drains in monkey and dog rooms be at least 6 in. (15.2 cm) in diameter and of the flushing type with special hair traps to avoid clogging.[126]

Although climatic conditions generally determine whether to provide cooling systems for human occupants, facilities for rodents always should be provided with complete air conditioning regardless of the geographic location of the facility.[2,26]

The size and type of air-conditioning system will depend on the locality, the available utilities, and the heating and cooling requirements of the facility. Individual controls should be installed for each animal room or zone to accommodate various animal species. Unless the institution provides night watchmen in the animal quarters, a dual thermostat arrangement or alarm system should be installed to safeguard against heating or cooling emergencies. Small animals that have been reared under ideal conditions with controlled temperature and relative humidity are very susceptible to temperature and moisture changes. If a power failure occurs and is not corrected within a short period of time, the breeding cycle of the animals will be upset or the whole colony may be lost.

Table 7 summarizes the recommendations for the optimum ranges of

TABLE 7. Summary of Ranges of Temperature and Relative Humidity

Species	*°C*	*°F*	*%RH*
Rat	18.3–22.8	65–73	45–55
Mouse	20–23.9	68–75	50–60
Guinea pig	18.3–23.9	65–75	45–55
Rabbit	15.6–23.9	60–75	40–45
Hamster	20–23.9	68–75	40–55
Dog	18.3–23.9	65–75	45–55
Cat	21.1–23.9	70–75	40–45
Monkey	16.7–29.4	62–85	40–75

relative humidity and temperature. It should be recognized that the ranges may be more critical for the small animals and for monkeys than for cats and dogs. Frequent hosing in dog and cat areas may cause high relative humidities and should be considered by the facility designer.

The architect and engineer should assure themselves that the scientific personnel who will use the facility have evaluated their criteria and selected optimum temperature and humidity range for each area. Since the figures in Table 7 are averages, specific applications may need different ranges. However, there is sufficient evidence to suspect that when temperature and/or humidity vary outside the stated ranges, the animals will be more susceptible to disease and will have a lower breeding rate.

In choosing the equipment to air-condition an animal facility, the architect/engineer should consider not only the BTU's generated by the animals at peak and normal activity ranges, but also the cfm required to keep the odor below an objectionable level. Table 8 summarizes useful design criteria.

It must be realized, however, that the data in Table 8 are based on frequent cage washing procedures and on normal animal holding densities. These procedures contribute to the reduction of objectionable odors, the control of enzootic conditions, and a reduction in required rate of air change. If the cages are to be washed more or less than the one-and-a-half to two times per week that Table 8 is based upon, a proportional adjustment should be made in the required cfm/animal. The graph, mentioned in footnote b, Table 8 will give heat emission for any animal whose weight is known. Therefore, if animals to be used vary greatly from weights listed, the BTU/hour/animal should be recalculated. Information concerning measurement of heat produced by various laboratory animals is available.[15]

Supply air for animal rooms ideally should be 100% outside air with no

TABLE 8. Criteria for Design of Mechanical Systems[15]

						Approximate Animal Heat Production per 228 sq ft Room Module				
						Total Heat		*Latent Heat*		
Animal	*Weight, grams*	*Basal Heat Production in 24 hr, kilocalories*	*Estimated Actual Heat Production in 24 hr,[a] kilocalories*	*Estimated Latent Heat Production in 24 hr, kilocalories*	*Number of Animals*	*Kilocalories per 24 hr*	*BTU[b] per hr (peak)*	*Kilocalories per 24 hr*	*BTU per hr (peak)*	*CFM/ Animal[c]*
Mouse	21	3.6	15	5	4200	63,000	21,000	21,000	7,000	.147
Hamster	118	10.0	40	8	2880	115,000	38,000	23,000	7,700	.406
Rat	300	27.0	100	35	960	96,000	32,000	34,000	11,000	.8150
Guinea pig	410	35.1	140	45	840	118,000	39,000	38,000	12,500	1.1550
Rabbit	2,600	117.0	235	55	72	17,000	5,600	4,000	1,300	12.0
Cat	3,000	152.0	300	75	36	11,000	3,600	2,700	900	12.0
Monkey	4,200	207.0	800	275	36	29,000	9,600	10,000	3,300	18.0
Dog	16,000	530.0	1000	250	12	12,000	4,000	3,000	1,000	90.0
Goat	36,000	800.0	1600	400	12	19,000	6,500	4,800	1,600	—
Sheep	45,000	1160.0	2300	600	12	28,000	9,000	7,200	2,300	—
Pig	250,000	4350.0	8700	2200	3	26,000	8,500	6,600	2,200	—
Pigeon	275	28.0	55	9	840	46,000	15,000	7,500	2,500	—
Chicken	2,100	114.0	225	36	420	95,000	32,000	15,000	5,000	10.5
Man	70,000	1650.0	6500	2200	2/3[d]	4,500	750	1,500	250	—

[a] Estimated on the basis that some animals under laboratory conditions liberate heat at four times the basal rate, others at only twice the basal rate as an average.

[b] See graph, page 1-163, paragraph entitled Heat Emission of Animals, in "Handbook of Airconditioning, Heating and Ventilating," Industrial Press, New York, 1959.

[c] Computed by multiplying average weights by cubic feet per minute (CFM)/pound (from Heating, Piping and Air Conditioning **10**:289–291, April 1938, and U.S. National Institutes of Health, Environmental Services Branch, Survey of animal rooms, Clinical Center, unpublished report, March 1956).

[d] One-third time for each of two workers, as an average.

recirculation. If air is recirculated, more efficient filtration for removal of odors and contaminants will be required. Odor removal is more difficult than removal of particulate matter. The outside supply air will not need sterilization, but will generally need filtering; however, provision should be made, if possible, to add high-efficiency filters at a later date if the required experimental conditions are revised. If the animal quarters are contiguous with the other research activities, the air supply system should be separate from other parts of the building. The air should, if possible, be introduced into the rooms near the ceiling and exhausted near the floor to remove the heavier-than-air ammonia fumes. The supply of air to the animal rooms should be introduced into the rooms at as low a velocity as possible, because higher-velocity air tends to increase the chance of the animals' catching disease.

In some installations, the exhaust ducts from animal rooms in which there is loose hair present have no grills over the exhaust duct. This allows hair to be drawn up the duct and blown out. If conditions of the surrounding area do not allow hair to be exhausted, then a filter should be used at the room exhaust register inside the exhaust duct. If this is not originally installed, space should be provided for installation at a later date. The exhaust from the animal portion of the building should be separate from the main exhaust header to prevent contamination of laboratory space in case of a power failure. The exhaust outlets should be located so as not to interfere or cause objectionable odors in critical areas. Fresh air intakes should be located above ground level on the prevailing wind side of the building and as far away as possible from any exhaust in order to prevent reentry of exhausted air.

Mechanization. Much of the cost of caring for animals is the labor and delay in cleaning or repairing cages. Industrial engineering methods can be used to advantage to study present or proposed operational procedures and to identify areas in which it would be most advantageous to automate equipment and material movement. The architect and engineer, when designing an animal facility, should attempt to incorporate as many labor-saving devices as are economically and practically feasible.

One of the newer innovations for automated operations, illustrated in Figure 28, is the vacuum extractor used for removing waste material from cages. This system may be used in several types of facilities and at different points in the equipment processing procedure. In conventional colonies, dirty cages may be removed from the animal room and brought to the cage washing area. The vacuum system can then be used as a means of removing waste from the cages and transporting it from the cage washing area to a storage bin or incinerator where it is destroyed. The vacuum system may

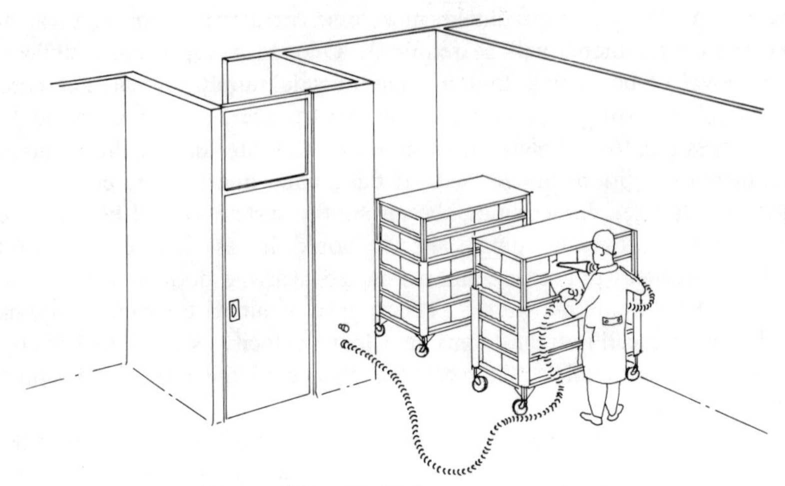

Figure 28. Built-in vacuum extractor.

also be used to clean cages within the animal room, as well as the room itself, thus eliminating the need to transport soiled bedding and animal waste through the building corridors. A large-diameter vacuum hose has even been used to remove mouse cages from a colony room directly to an incinerator. The use of the vacuum extractor in highly infectious animal facilities is not recommended.

In cage washing areas, the operations of removing soiled bedding, cage washing, water-bottle washing and refilling, replacing cage bedding, and fresh food supply can be integrated so that all functions are carried out progressively with a minimum of manual labor. Automatic cage and rack washers can operate at much higher water temperatures than human hands can stand, and therefore can kill a large percentage of the microorganisms on the equipment. The cages and racks are also washed faster, and since the machines are automatic, there is no question of variation from the set standard of cleanliness. Using the vacuum extractor at the beginning of the assembly line to remove soiled bedding and an automatic bedding dispenser at the clean end allows minimum handling of the cage. Machines have been developed to perform the complete process of water-bottle cleaning and filling. The automatic washers clean and sterilize the bottles more thoroughly because they are designed specifically for this operation. An important consideration in the design of washing facilities is provision of adequate space for spare cages, both clean and dirty.

The addition of an efficient incinerator adjacent to the animal facility will eliminate the cost of removing waste material. There will be less chance of spreading contamination outside the facility grounds if the diseased material is destroyed in the incinerator.

In any operation, the equipment handling and service area should be centralized in the animal colony. If expansion is planned or is possible, a location should be selected for the central services so that the area will not be isolated or inconvenient if expansion or extension does take place.

Other areas in which automation has either been used or is thought to be feasible are: automatic watering devices for animal cages; automatic flush pan to catch droppings; overhead storage of bedding for automatic dispensing; specially treated corrugated paper as a cage liner; and an automatic weight sorter for separating animals in different weight ranges. As more experience is gained in the field of animal care, many different labor-saving devices undoubtedly will be developed. Although the architect/engineering ordinarily cannot be expected to develop these devices, he should contact groups who have had experience in these areas, or employ knowledgeable consultants.

Rooms for Handling Radioactive Materials. Just as most biological laboratory directors wish to have one or more rooms equipped for chemistry, they now wish to have one or more laboratories equipped to handle radionuclides. Federal, state, and local regulations are specific for shielding, personnel protection, and disposal of radioactive wastes. A list of pertinent Federal Radiation Protection Regulations can be found in Appendix I. These design criteria therefore will not attempt to cover the specific requirements for radiation protection. It is most important that scientists and management determine the amount of radiobiology to be done prior to design of a building, since designs of these laboratories are significantly different, involve many federal regulations, and have higher construction costs. It may be difficult to meet radiological safety standards after the building is constructed. For example, if gamma-ray emitters are to be used, appropriate lead shielding must be provided and the weight of the shielding may exceed the original live load for which the floor was designed.

All surfaces in laboratories and animal rooms where radioactive materials are to be used must be monolithic, nonporous, and washable. Cracks, crevices, and joints must be sealed. Vinyl, asbestos, rubber, or linoleum sheets can be applied over a concrete floor to provide protection, since these materials are nonporous and can be fitted if necessary for radiological decontamination. Epoxy resin paints or polyurethane coatings will seal plaster walls effectively if properly applied. Stainless steel sinks, bench tops, animal cages, and other items that might be contaminated with radioactive materials are recommended, since they can easily be decontaminated.

Provisions should be made to collect, store, and monitor all liquid and solid wastes from radiological areas. All drains from radiological work areas should empty into buckets or other suitable containers. When the radiation levels of waste materials drop to certain specified levels, the materials may be disposed of by pouring liquids into the sanitary sewage system, by incineration, or by burial of solids. However, if the radiation levels are above those specified, the materials, including animal carcasses, must be taken to an authorized radiation disposal center.

A special fume hood is required for all radiological procedures in which there is any chance of air pollution by radioisotopes. The air flow rate into the hood opening should be at least 100 linear ft per min but should not exceed 200 ft per min. The hood should be equipped with an ultra-high-efficiency filter for exhaust air.

In the microbiological containment facility the design engineer may find that he must provide facilities for infectious microbial agents tagged with radioactive materials. A stainless steel bacteriological cabinet with an open front or a closed, gastight cabinet with an ultrahigh-efficiency filter is suitable for controlling both hazards. Whether the open front, partial barrier cabinet or the gastight (absolute barrier) cabinet is used will depend on the microbiological safety requirements; either will be satisfactory for the radiological hazard.

In some instances when highly radioactive materials are to be used, the design engineer will have to plan on installing remote handling equipment such as tongs, forceps, clamps, or remote, mechanical hand systems. The latter systems are expensive and ordinarily will require large amounts of shielding.[81]

Rooms for the Electron Microscope. Many containment facilities will require an electron microscope. A complete electron microscopy laboratory must provide space and facilities for operation of the microscope and its accessories, as well as laboratory facilities for specimen preparation, and photographic facilities for developing and printing.[114] The room should be located for easy elimination of excessive vibration. This may be generally recognized when vibration levels in the region of 10^{-4} "g" are met or exceeded.[114] Basements or first floors are usually more stable than upper floors. Elevators, fans, pumps, and other motors can not only cause excessive vibration, but electrical equipment may generate troublesome stray magnetic fields. The electron microscopy room should have at least high-efficiency filtration of supply air to control dust. Large fluctuations in temperature and relative humidity should be avoided. Ordinarily the room should be at a positive air pressure to prevent entrance of contaminants. The heat load imposed by the microscope and its power supply should be considered when planning temperature control. The usual required utilities are water, electricity, and compressed air.

Walk-in Refrigerator and Incubator Rooms. More space is usually required for refrigeration than for incubation. Commercial prefabricated units are usually suitable for most needs.

Provisions should be made in the design of the building for condensate drip lines to run from the cooling coils through the floor rather than on top of the floor. All crevices between the refrigerator or incubator, and the walls and floor of the room, should be sealed unless the unit is readily movable. Walk-in refrigerators, and usually walk-in incubators, should be provided with floor drains to facilitate cleaning operations. The floor drains should be placed as near the back wall as possible. Traps should be deep enough to prevent their being emptied by positive pressure when the doors are closed. Doors to walk-in refrigerators and incubators should have a sealed, double-glass viewing panel in the center with the bottom of the viewing panel located approximately 58 in. above the floor. The hardware for walk-in refrigerator and incubator doors should be corrosion-resistant and should be equipped with a padlocking handle and an interior automatic release that operates whether or not the outside handles are padlocked or free.

Solvent Storage Room. One of the major causes of fires in research facilities is misuse in the storage and handling of flammable solvents. Most facilities should have a special storage room or cubicle for flammable materials. The flammable material storage room should conform to the National Board of Fire Underwriters (NBFU) Requirements for Type "B" inside storage or mixing rooms, and should be protected with a fixed carbon dioxide (CO_2) extinguisher system installed according to the National Board of Fire Underwriters or the National Fire Protection Association requirements. Approximately one pound of CO_2 is required for each 15 cu ft of space. The air exhaust fan for the flammable material storage room should be spark-resistant (AMCA Type B) and should have an explosion-proof motor.

Mobile Containment Facility. The establishment of a facility suited for research programs involving hazardous infectious agents frequently requires years. Providing these facilities has been made increasingly difficult by an acceleration in the pace of biomedical research which necessitates rapid changes in program emphasis and direction. In addition, research contractors and grantees are frequently unable to acquire adequate containment laboratories due to the long engineering and construction times and the high capital investment. As a result, the scientist most capable of conducting a particular investigation may not be the scientist with available, adequate laboratory facilities. Recently, a prototype mobile unit was designed, developed, and evaluated for the National Cancer Institute by the Dow Chemical Company.

The mobile containment laboratory was designed to have a high degree of mobility, to require little site preparation, and to be reasonably self-sufficient. Functionally, the laboratory had to include appropriate total containment equipment and accessories, space for housing small laboratory animals, and space for common nonhazardous laboratory functions. Materials of construction had to be easy to clean and resistant to decontaminating chemical and the anticipated stress of transportation. Common laboratory utilities and services were required.

The outside dimensions were set at 40 ft (12.19 m) long, 8 ft (2.44 m) wide, and 13 ft 6 in. (4.11 m) high for movement over the highway without special oversize permits, escorts, or night movement restrictions. Since the laboratory would incorporate heavy equipment, the light-framed mobile home type of unit was not suitable. The semitrailer van, as exemplified by the high-roof, low-floor moving van, was selected because it can carry the anticipated loads without significant distortion and has practically unlimited mobility.

One of the design requirements was provision of primary and secondary barrier systems. The primary barriers are represented by a gastight, in-line cabinet system, shown in Figures 29 and 30. It includes a necropsy hood, a refrigerated ultracentrifuge cabinet, two cabinet-enclosed, well-type incubators, and a general-purpose cabinet. The opposite side houses a fume hood that offers a partial containment capability for bench work and also contains a through-the-wall autoclave. The remainder of the laboratory is equipped with benches containing a drawer-type deep freeze, refrigerator, and storage units.

The secondary barrier system consists of impervious stainless steel walls and ceiling and the monolithic polyurethane floor coating. All joints are welded or sealed with epoxy or silicone compound and all interior angles smoothly coved for easy cleaning. The differential air pressure maintained in the various compartments assures a flow of air toward the potentially higher risk laboratory and animal room, thereby providing an additional secondary barrier.

Components necessary to provide full support to the laboratory functions are located below the personnel deck. Potable water from an outside source is stored in two 125-gal tanks and is moved under air pressure to the laboratory space above. An air compressor supplies pressure to the water tanks,

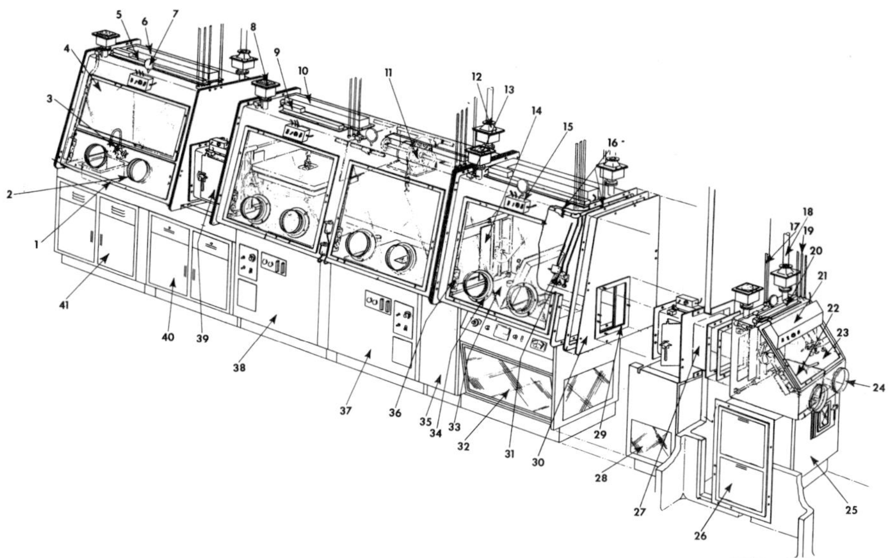

Figure 29. Biological safety cabinets and components—Class III. (*Courtesy S. Blickman, Inc.*)

1. Stainless steel sink 8″ × 10″ × 6″.
2. Stainless steel removable gasketed glove port panel.
3. Hot and cold water gooseneck mixing faucet.
4. Bolted safety glass viewing panel.
5. Stainless steel UV light housing.
6. Stainless steel fluorescent light housing.
7. Magnehelic gauge.
8. Stainless steel removable inlet absolute filter housing.
9. Stainless steel UV light housing.
10. Stainless steel fluorescent light housing.
11. Interior mounted removable automatic hoist lift for lazy susan with exterior controls.
12. Stainless steel removable exhaust absolute filter housing.
13. Feed thru connection for interior duplex receptacle.
14. Stainless steel removable horizontal sliding door for 12″ × 16″ access opening.
15. Stainless steel electrical control box with power supply cord and cap.
16. Interior stainless steel decon spray system.

17, 18, 19. External plumbing and ventilation connections.

20. Stainless steel UV light housing.
21. Stainless steel hinged fluorescent light housing with electrical control switches and power supply cord and cap.
22. Stainless steel interior of cabinet air lock door for 10″ × 12″ access opening.
23. Removable bolted safety glass viewing panel.
24. Stainless steel AEC-NBL type glove ports.
25. Stainless steel front facia panel with exterior of air lock access door with viewing window.
26. Stainless steel removable cabinet support and storage compartment.
27. Stainless steel removable 12″ × 16″ × 26″ long air lock with front loading 12″ × 14″ opening access door and thru wall sealing flange.
28. Stainless steel front facia panel with ventilation screen for centrifuge.
29. Neoprene air lock gasket.
30. Stainless steel 7 GA. adaptor panel with interior of cabinet horizontal sliding stainless steel air lock access door.
31. Stainless steel service bracket with stainless steel air, gas, and vacuum valves with serrated nozzles.
32. Bottom mounted refrigerator centrifuge with interior of enclosure mounted centrifuge access door.
33. Bolted safety glass viewing window.
34. Window mounted stainless steel AEC-NBL type glove ports.
35. Stainless steel front facia panel.
36. Interior stainless steel duplex receptacle housing.

37, 38. Stainless steel front facia panel with incubator controls, electrical access panel, and CO_2/ air flow meter.

39. Removable stainless steel 12″ × 16″ × 16″ long air lock with front loading 12″ × 14″ access opening.
40. Stainless steel base support cabinet with drawers and doors.
41. Stainless steel base support cabinet with doors.

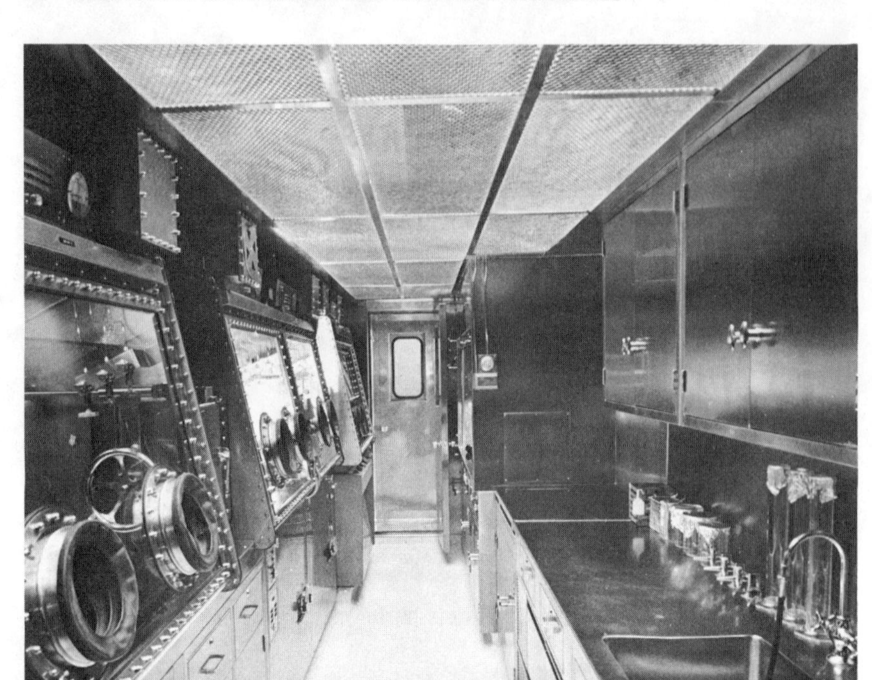

Figure 30. Biological safety cabinets and components—Class III. (*Courtesy S. Blickman, Inc.*)

the pneumatic ventilation controls, and the laboratory air system. A vacuum pump system services the laboratory and the autoclave. Four 20-lb carbon dioxide bottles on a manifold system with automatic pressure control supply the CO_2 incubators. Fuel gas for laboratory burners is provided by a 20-lb propane bottle attached externally to the trailer.

Steam for humidification of room air supply and formaldehyde/steam decontaminating of the safety cabinets is generated by an electrically heated boiler. Electricity also heats water for hot water pipes to the laboratory and shower. The equipment deck includes an emergency lighting unit that will automatically illuminate auxiliary lamps in the laboratory and animal rooms in the event of a power failure. A small refrigeration unit and circulating pump supply cooling water to the centrifuge on the deck above.

A large portion of the equipment deck is occupied by the sewage treatment system. Sewage from the personnel deck is collected and released in 4½-gal batches into a chlorine contact tank. Concentrated chlorine is metered simultaneously into the contact tank from a chlorine solution tank. The chlorine-treated sewage is then pumped into a 250-gal holding tank where the day's accumulation is held overnight before disposal.

One hundred per cent replacement room air is supplied to the mobile unit. This air is fully conditioned and temperature and humidity are controlled from within the laboratory. The air is drawn in at a rate of 1150 cfm, heated if necessary, cooled by refrigeration coils, and passed through absolute filters before distribution to the various compartments. The distribution ducts leading from the main plenum contain heating components and are located above the ceiling of the personnel deck. There are also seven exhaust blowers occupying this overhead space that remove air from the safety cabinetry and the personnel compartments. All of the exhaust air passes through absolute filters before release. The ventilation system is designed to provide about 20 changes per hour in the laboratory and animal rooms.

A small compartment accessible only from the exterior is located at the forward end of the trailer. This contains the electrical distribution panels and pneumatic control system for the ventilation and air-conditioning controls. The unit is equipped with a 150-KVA stepdown transformer so that it is adaptable to any suitable 440-V or 220-V power source.

Upon completion, such factors as the air handling system, gastight cabinetry, and laboratory services and equipment were evaluated for containment efficiency and specified performance. The gastight safety cabinets and pass boxes were leak-checked using a halogen leak test with dichlorodifluoromethane at 3-in. water gauge pressure.

The HEPA exhaust filters were checked for conformance to the specification after installation using a DOP generator and detector.

The liquid waste inactivation system was calibrated to determine the relationship between final residual chlorine concentration and pumping rate, pumping time, and chlorine feed solution concentration. Complete inactivation of *Bacillus subtilis* spores was obtained after a 60-min hold period in four trials with a final concentration of 350 ppm and waste influent flowing at a rate of $\frac{3}{4}$ gal per minute.

A prototype of a mobile virus research laboratory shown in Figure 31 has been designed, fabricated, and tested. It has been use-tested in the field at the Communicable Disease Center in Atlanta, Georgia. Further evaluation and development is being performed by the National Cancer Institute.

Microbiological Barriers

The Building. The primary-secondary microbiological barrier concept[101] provides a clear and simplified explanation of microbiological barriers. Enclosures, barriers, or other containment devices that immedi-

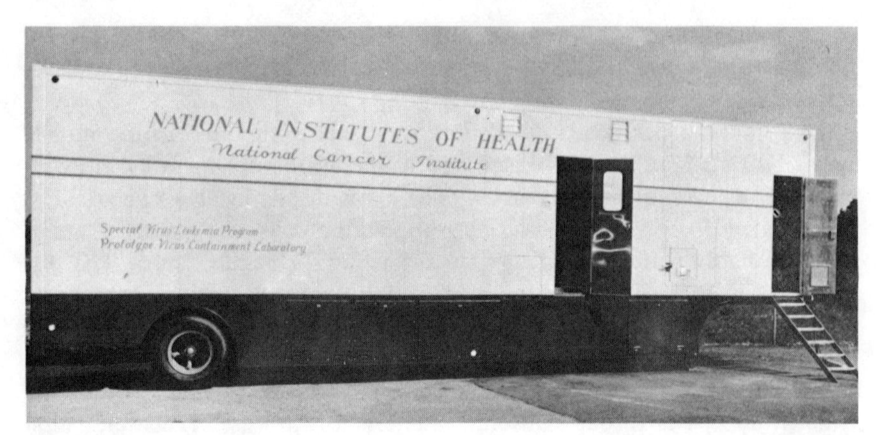

Figure 31. Exterior view, mobile containment laboratory. (*Courtesy National Cancer Institute.*)

ately surround or enclose infectious materials or materials to be protected from contamination are designated as primary barriers.

The secondary barriers in a facility are the features of the building that surround the primary barriers. These provide a separation between the containment areas in the building and the outside community and between different individual areas within the same building. These secondary barriers provide supplementary microbiological containment, serving mainly to prevent the escape of microorganisms if and when a failure occurs in the primary barriers and to limit entrance of microbial contaminants that would be available to enter primary barriers. For further discussion and examples of primary and secondary barriers see *Facility Functions,* page 39.

Room Arrangements. The rooms within a building should serve to separate areas of varying degrees of contamination or cleanliness. Most facilities will utilize part of the building for offices, library, conference room, lunchroom, etc. Traffic usually should move from the administrative area through a change-room system into the operational area. In the infectious disease laboratory, personnel leaving the laboratory area should discard their laboratory clothing into an ultraviolet-protected clothes hamper and shower before moving to the clean side of the change room. In the product protection type of facility, personnel should shower or pass through an airwash upon entering the laboratory area, but ordinarily no precautions need be taken as the workers leave. For personnel protection, rooms should be arranged in a sequence so that traffic moves from areas of lesser to greater potential hazard or exposure, whereas for product protection, traffic should move from area of greater contamination to areas of lesser contamination.

In some instances there may be a requirement for rooms within rooms. A small room within a larger room may be a clean room for operations such as tissue culture work done within a laminar air flow bench. In this case, the inner room would be maintained under a positive air pressure. On the other hand, the inner room may contain a bacteriological work cabinet where hazardous infectious disease work is carried out. In this instance, the inner room would be maintained under a negative air pressure relative to the main laboratory and the cabinet would be at a negative presssure to the inner room.

The most important mechanism for preventing the escape of hazardous particles from an area or penetration into an area is through controlled, directional flows of air. In the infectious disease laboratory, the air should flow, due to a series of pressure differentials, from the clean office areas, through the change rooms, toward the most hazardous areas. In the laboratory areas, air should be supplied to the corridors, but not exhausted from the corridor. Air flows should be from the corridor into each of the rooms, thus isolating each room.

Ventilated Cabinets and Hoods. The ventilated cabinet is one of the most important primary barriers available to the microbiologist for isolation and containment of his work and for keeping the laboratory environment, the building, and the surrounding area free of contamination. All procedures capable of generating infectious aerosols should be carried out in the ventilated cabinet. There are several styles of ventilated cabinets available for use (see Figures 32, 33). The single unit, open or closed front cabinet is usually referred to as a partial barrier ventilated cabinet. This cabinet can be used with the glove panel removed, depending upon an inward flow of air of at least 100 linear ft per min to prevent escape of airborne particles. Alternatively, it can be used with the glove panel fastened in place and arm-length gloves attached. In this case, it will be maintained under a reduced air pressure of about 1 in. of water gauge. When operated closed, the partial barrier cabinet needs an attached air lock for movement of materials. A third mode of operations consists of use of a cabinet with glove panel attached, but with the gloves removed.

The second type of ventilated cabinet is the gastight cabinet system, referred to as an absolute barrier cabinet. Absolute barrier cabinets are connected to form a modular cabinet system with enclosed refrigerators, incubators, deep freezers, animal holding areas, autopsy cabinets, and any other special cabinets that might be needed for conduct of research behind a barrier. Figure 34 shows the interior of a cabinet with a back-mounted

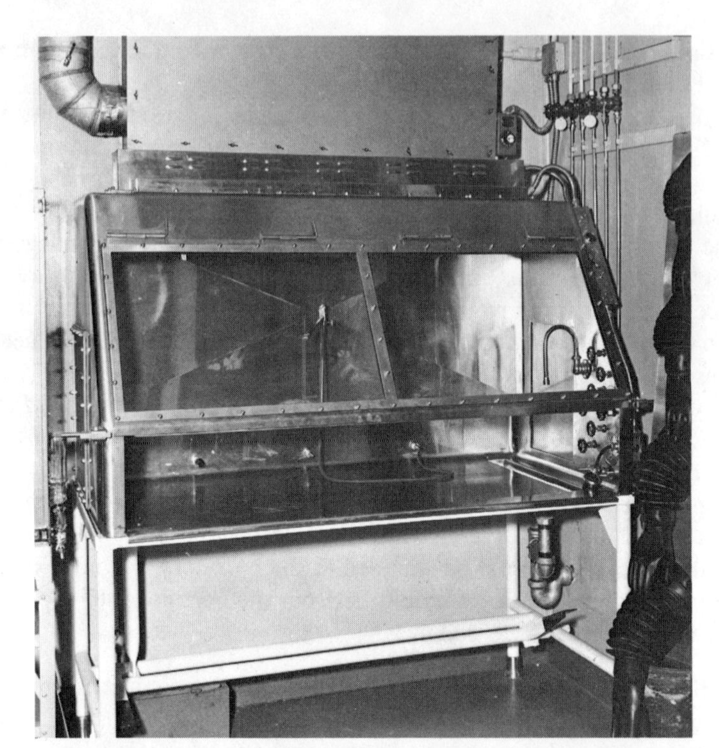

Figure 32. Open-face biological safety cabinet. (*U.S. Army photograph.*)

Figure 33. Modular gastight cabinet system. (*U.S. Army photograph.*)

refrigerator. Air is drawn into the cabinet system through ultrahigh-efficiency filters and is exhausted through ultrahigh-efficiency filters. If needed, the exhaust air may also be passed through an air incinerator. The gastight cabinet is tested for tightness by injecting chlorofluorohydrocarbon (Freon) gas into the cabinet to create 2–6 in. water gauge of positive air pressure within the cabinet. The presence of this gas outside the cabinet as shown by a G.E. Type 2, Halogen Leak Detector indicates a leak which must be sealed. Because the cabinet system is operated under a reduced air pressure of about one inch of water, the safety factor involved is very large.

Reduced-pressure enclosures can be of any size or shape to house a specific function or piece of equipment. Temporary, inexpensive enclosures[93] can be quickly and easily fabricated from plastic. Such enclosures, made of flexible film supported by aluminum tubing, can be custom-made

Figure 34. Absolute barrier safety cabinet equipped with refrigerated centrifuge. (*Courtesy S. Blickman, Inc.*)

to contain various types of equipment such as microscopes and balances. These devices are useful to initiate experiments involving potential hazards and used until the research effort is either abandoned or proves desirable, at which time permanent partial or absolute barrier systems or cabinets can be employed.

Laminar Flow Cabinets. Prototype laminar air flow units were first designed in 1961 (Whitfield, 1962). Laminar air flow was immediately recognized as valuable for controlled environmental work areas in the aerospace industry, particularly in the manufacture and assembly of high-precision electronic components where the slightest trace of dust or particulate contamination could cause malfunction.

Almost from the beginning of the laminar flow era, it was apparent that many areas of application existed in the biomedical field. Laminar flow devices have found application in the hospital, both in surgical theaters and in patient rooms, in the microbiological assay of space hardware, and during the preparation of tissue cultures. Laminar air flow has found wide use in sterility testing procedures and in the production, filling, and packaging of drugs and pharmaceuticals. Still another use is during the rearing or holding of specific-pathogen-free or defined-flora animals or for protecting normal animals under test from zoonotic disease.

While there is no doubt of the value of laminar air flow for providing better control of the microbial contamination in many diverse areas, it is important that proper attention be paid to its manner of use. Careful planning for the placement of equipment and supplies and control of the movement of people and objects in the laminar air stream is necessary. It is important to realize that equipment or objects closest to the supply filter wall will have the greatest degree of biological protection, while objects further downstream may not be as well protected. Environmental spaces protected with laminar air flow should contain only the bare minimum of laboratory equipment and supplies because cluttering of the work area will obviously disrupt laminar flow patterns. An important point to stress in relation to the above comments is that laminar air flow devices provide control over airborne particulate contamination only and will not remove surface contamination. However, unless airborne contamination is controlled, exposed surfaces will become contaminated because surface particle collection always occurs when airborne particles are present.

The various laminar air flow devices utilize either the cross-flow or the downflow principle. The cross-flow principle is illustrated by the laminar flow clean bench shown in Figure 35. This type of device is used for product or process protection in many industrial firms and in the drug and medical fields.

The laminar downflow principle is illustrated by the curtained unit shown in Figure 35. While permanent-type installations with a grated floor for the air return would be recommended for many applications, the device illustrated in Figure 35 makes possible the use of portable or temporary clean room areas for microbiological environmental control. A recent publication by McDade, et al. (1966b) provides a microbiological evaluation of several laminar flow installations. At an invited session at the 67th Annual Meeting of the American Society for Microbiology, seven papers on the use of laminar air flow for controlling microbial contamination were presented.

Ultraviolet Barriers. Ultraviolet can be used effectively as a microbiological barrier to separate areas of different levels of contamination.[89,97,133] It can be used in air locks and as a screen across the tops of animal cages (see Figure 23). The effective use of this radiation as a microbiological barrier requires an understanding of its effectiveness as well as its limitations. The 2537 Å wavelength emitted by the mercury vapor germicidal lamp has limited penetrating ability and is effective primarily on exposed surfaces or in air. Proper intensity, exposure time, and lamp maintenance are critical. The use of UV in the microbiological laboratory is fully detailed in several recent publications.[97,133] The pharmaceutical and other industries use UV as a microbiological barrier for product protection.

Air–Supplied Suits. In several situations, the ventilated suit (see Figure 26) is used as a microbiological barrier[12] in the infectious disease laboratory, in facilities for sterile assembly of spacecraft, in the production of germfree animals, and in the treatment of low-resistant patients. In large rooms for animals exposed to organisms pathogenic for man, it may be faster and safer for the animal caretaker to wear a ventilated suit than to attempt to isolate each infectious animal. The animals can then be housed in open cages rather than individually ventilated cages. In germfree application, the opposite goal of protecting the animals can also be met with ventilated suits. If the ventilated suit system is used in any facility involving infectious organisms, a disinfectant shower room is required for decontamination of the suit before the wearer leaves the area.

Respiratory Protection. Inhalation of accidentally produced infectious aerosols is the most frequent mode of laboratory infection.[99] Therefore, when laboratory personnel are exposed to procedures likely to create aerosols, respiratory protective devices should be used. A variety of respirators (see Figure 24) and air-supplied head hoods (see Figure 25) are available. The air-supplied head hood is much more comfortable to

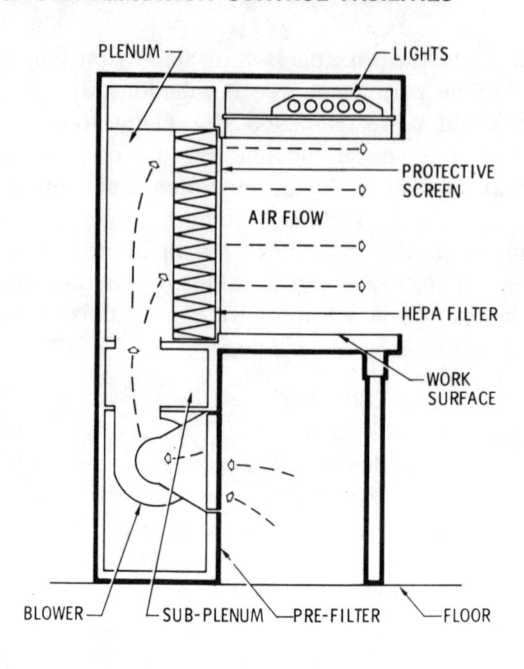

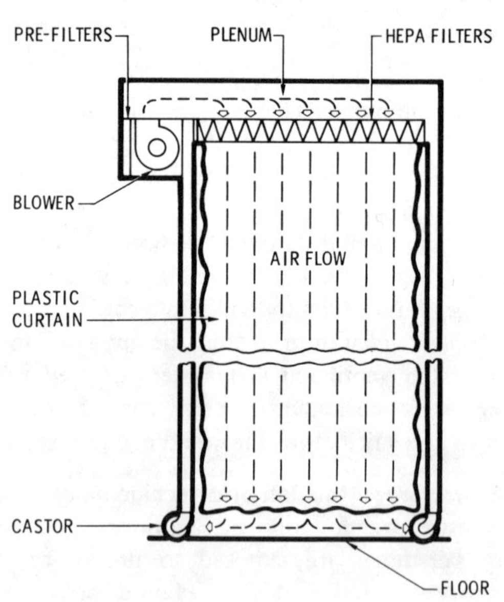

Figure 35. Upper: Cross section of a laminar flow clean bench. Lower: Curtained downflow laminar flow unit.

wear over a period of several hours than is the respirator; however, the ventilated head hood requires a rather elaborate and expensive air supply system.

Disinfectant Foot Baths. Areas such as the doorway from an animal room, the entrance to a clean assembly area, or an isolation facility for treatment of low-resistant patients can use a tray containing a disinfectant. The disinfectant used will vary with type of agent under study in the facility. Spongy or tacky mats are frequently used at the entrance to laminar flow rooms, primarily as a means of reducing particulate contamination.

Heat and Gaseous Decontaminants. Heat is the most effective and reliable method of inactivating microorganisms and should be used whenever possible. The exposure temperatures and times required for sterilizing are known and can be really controlled. Dry-heat ovens containing air or an inert gas can be used while passing some materials and supplies in and out of sterile barrier systems. Steam sterilizers also are recommended for passing materials in and out of barrier systems; at comparable temperatures, moist heat is faster than dry heat and more reliable for surface decontamination.[103] Heat sterilization of entire barrier systems is being considered for spacecraft sterilization facilities.

The double-door, through-the-wall autoclave, extending from the contaminated side of the building to the clean receiving room, is one of the microbiological barriers used to prevent contaminated materials from reaching the outside of the building or to prevent contamination from reaching a protected area. All materials, equipment, trash, etc. leaving a building where pathogens are used should be sterilized as they are passed out.

Ethylene oxide (ETO)[91,92] is a chemical used in a gaseous form. When employed in closed systems and under controlled conditions, excellent decontamination or sterility can be achieved. However, the properties and limitations should be thoroughly understood in relation to the barrier system.

Ethylene oxide* is a highly penetrating and effective sterilizing gas, convenient to use, versatile, noncorrosive, and effective at room temperature. However, the gas is slow in killing microorganisms and usually must be mixed with other gases to avoid explosion hazards. Ethylene oxide is widely used to treat many items that cannot tolerate heat sterilization. It

*Ethylene Oxide Mixture, Pennsylvania Engineering Co., Philadelphia, Pa.; Cryoxcide American Sterilizer Co., Erie, Pa.; Steroxcide, Wilmot Castle Co., Rochester, N. Y.

has been used is mixtures with carbon dioxide or nitrogen, which requires that it be used under pressure. Its most extensive use today is in the form of a low-pressure mixture with chlorofluorohydrocarbons (Freons) in disposable cans or cylinders. In this form it is a highly practical and convenient tool for increasing the usefulness of the laboratory autoclave. A steam autoclave can be converted to its use without interfering with the use of the autoclave with steam.[110] Definite limitations to the use of ethylene oxide are the required exposure time and the fact that the efficiency of the gas is a function of the humidity. In practical, nonflammable concentrations and at room temperature, a minimum of six hours is required to sterilize materials contaminated with bacterial spores. Longer (overnight) exposures are recommended for routine use. Another limitation is that neoprene gloves, clothing, footwear, or other plastic, rubber, or leather wearing apparel that have been treated with ethylene oxide must be thoroughly aired for 24 hrs or more before use in contact with the skin to avoid the irritating action of absorbed ethylene oxide on human tissues. The effect of moisture on ethylene oxide sterilization also should be recognized by any potential user.[44] Ethylene oxide gas mixtures can be used to sterilize microbiological barriers prior to use or to treat certain materials passed into or out of the barrier. Table 9 summarizes information concerning the above two decontaminants and other agents, gives recommended treatments, and cites some limitations in their use.

Communication Through Barriers. Communication between the areas with different levels of contamination in a facility can be accomplished by several methods. They all serve to reduce traffic into and out of protected areas and thereby reduce the chances of contamination breaching the barrier. The easiest and most inexpensive method uses clear plastic (Saran or Mylar) speaking diaphragms and viewing panels. Doors opening into laboratories, animal rooms, air locks, clean rooms, or any area where a different level of hazard or contamination exists should be equipped with a speaking diaphragm and a viewing panel. Large glass panels and several speaking diaphragms should be located in the infectious disease facility between clean and contaminated offices. Effective speaking diaphragms assemblies are available commercially.

In some instances, for example when operators of gloved safety cabinets must communicate with different areas of the building, electronic paging telephone communication may be desirable. Such a system should also allow local phone calls to be placed from inside the contained space. Another concept which is currently being investigated is the use of small television cameras for monitoring specially contained areas where there is a danger of accidents, or for monitoring control panels from a remote

TABLE 9. Some Recommended Agents and Treatments for Sterilization or Decontamination in Microbiological Barriers

Sterilization of Decontamination Agent	*Recommended Treatments*[a]	*Limitations of Use*
Moist heat (autoclave, high vacuum)	135° C, 3–5 min	Not including come-up time. Effective if material is pervious to steam; otherwise effect is essentially that of dry heat.
Moist heat (autoclave, no vacuum)	121° C, 15–30 min	Not including come-up time or size of vessels.
Dry heat	160° C, 2 hrs; 170° C, 1 hr	Not including come-up time. Other combinations of temperature and time are acceptable.
Ethylene oxide gas (in a nonexplosive gas mixture)	25°–55° C, 300 mg per l., 6–16 hrs, 30%–60% RH	Will not penetrate solids. Adsorbed in rubber and plastics necessitating aeration if material is to contact skin.
Peracetic acid spray	25° C, 2% with 0.1% surfactant, continuous for 20 min	Corrodes many metals. Degrades to active acid, oxygen, and water.
Steam-formaldehyde vapor	25° C, 1 ml per cu ft in air, RH above 80%, 30 min (cabinets) or 10 hrs (rooms)	Formaldehyde polymerizes on surfaces often necessitating long aeration periods prior to re-entry.
Beta-propiolactone vapor	25° C, 200 mg per cu ft in air with RH above 70%, 30 min (cabinets) or 2 hrs (rooms)	Aeration required prior to entrance into area.

TABLE 9—Continued

Sterilization of Decontamination Agent	*Recommended Treatments*[a]	*Limitations of Use*
Dunk-bath formalin (37% HCHO)	25° C, 10%, 10 min	Irritating fumes.
Sodium hypochlorite solutions	25° C, 500–5000 ppm with 1% surfactant, 5 min	Corrodes many metals.
Strong tincture of iodine (%)	25° C, 10 min	Stains many materials.
Ultraviolet radiation	25° C, 800 micro-watt min per sq cm	Low penetrating power. Use limited to clean exposed surfaces and air. Bulbs must be checked and kept clean.

[a]Recommendations are based on maximum effectiveness against bacterial spores and are made on the basis of direct or conservatively extrapolated experimental data.

location. These devices are also employed for monitoring patients under extreme isolation techniques such as the Life Island.

Arrangements must also be made for transition of materials and small supplies, such as books, data sheets, and similar items, between clean offices and offices or laboratories in the contained area. Typically, this may include a small through-the-wall ethylene oxide gas chamber for the cold sterilization of heat-sensitive materials[92] and a UV apparatus for decontaminating single sheets of paper passed out of the contained area.[95]

Disinfectant Dunk Tank. The disinfectant dunk tank allows quick, convenient passage of items into and out of an enclosure, such as a gastight cabinet system, without the transfer of air or microorganisms. A baffle extends from the interior of the enclosure to a level several inches below the surface of the disinfectant so that items passed into or out of the enclosure are completely submerged as they are passed below the baffle (see Figure 36). The standard disinfectant dunk bath or tank designed for use with the gastight cabinet system is mounted so that the tank can be rolled under the cabinets when not in use.

Filters and Air Incinerators as Microbiological Barriers. (See "Heating, Ventilating, and Air Conditioning," page 117.)

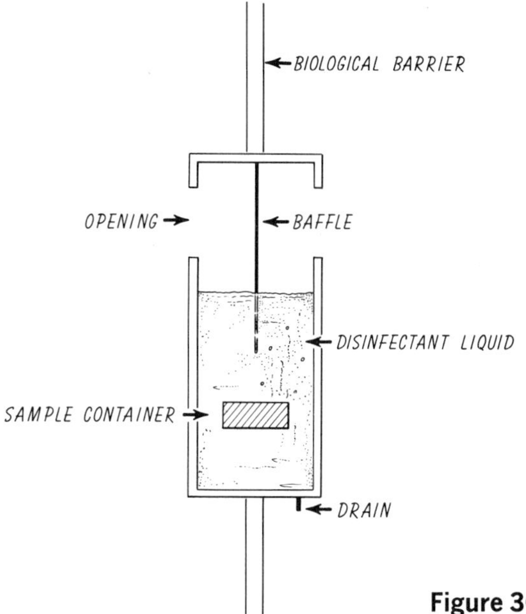

Figure 36. Disinfectant dunk tank.

Special Design Details

The following are a number of building and room features that either are found exclusively in microbial containment facilities or are of unusual importance to these facilities.

Rodent- and Vermin-proofing. All microbiological containment facilities should be completely rodent-proofed in accordance with the criteria established by the U. S. Public Health Service in its book entitled "Rodent Borne Disease Control Through Rodent Stoppage" (1961). This book provides detailed design information that will assist the architect engineer in rodent-proofing structures, especially animal rooms and storage areas for animal food and bedding. These spaces are particularly susceptible to infestation of insects. Additional material on rodent- and vermin-proofing can be found in several other books.[73,77] If research conducted in an infectious disease facility is with organisms pathogenic or infectious for feral rodents and vermin, special care must be taken to prevent contamination of these animals.

Fire-resistant Construction. The cost of total fireproof construction is so great that, in general, it is difficult to justify for most microbiological containment facilities. However, fire-resistant materials and construction should be used to the fullest extent practicable. Wood and other combustible materials should be avoided when possible. Certain areas, such as chemical storage areas, can be protected by CO_2 extinguisher systems. The use of sprinklers in some biologically hot areas has been discouraged, since the infectious material could be distributed by the water spray.

Waterproofing. The floors in both contaminated and noncontaminated areas that are located above occupied spaces and subject to washdowns should be of waterproof construction. In containment areas, concrete floor slabs on grade should be placed on a continuous polyvinyl chloride sheet of 6 mil minimum thickness, extended up the wall for 6 in., with sleeves of the same material around all penetrations. The sleeves should be sealed to the floor membrane by heat or with solvents. Concrete curbs, poured integral with the floor and at least 4 in. high, should be installed under walls; at door openings of solvent storage rooms, shower stalls, and drying spaces; air handling plenum chambers and animal rooms; and under the walls separating noncontaminated from contaminated areas. The face of the curb should be flush with the face of the wall above. Expansion joints in the floor should be provided with a continuous water stop, and should be filled to a depth of $\frac{1}{2}$ in. with a two-component, polysulfide sealant,* or a one-part silicone sealant.† A watertight seal should be provided at each floor level around all pipes, conduits, instrument tubing, and ducts whenever they pass through the floor, wall, or ceiling.

Floor Finishes. A variety of floor finishes is currently available. The utility of the finishes and the unit cost vary over a wide range. Obviously the architect/engineer should endeavor to select the finish most suitable for the particular use and conditions to be encountered. Tile and sheet linoleum should be avoided for rooms that are to be washed frequently with disinfectants and water. A satisfactory floor finish for many purposes can be obtained by using a nonslip metallic aggregate top layer for concrete that is steel-troweled to a glass finish. In this process, it is necessary to treat the uncured surface with a concrete hardener, seal it, and finally finish it with a nonskid wax. Such a floor will withstand almost any use

*(1) Products Research Corp.—PRC 1422—Class B-2 (2919 Empire Avenue, Burbank, California).

(2) Minnesota Mining & Manufacturing Co.—EC 1675—Class B-2 (900 Bush Avenue, St. Paul, Minnesota).

(3) Pro-Seal & Manufacturing Co.—Pro-Seal 890—Class B-2 (19451 Susana Road, Compton, California).

†(1) Dow-Corning Silicone Sealant, Dow-Corning Co., Midland, Michigan.

(2) RTV-106, General Electric Silicone Products Dept., Waterford, New York.

or abuse, from flooding with disinfectants to the heavy traffic of carts and handtrucks.

A satisfactory floor finish also can be obtained by applying an epoxy-aggregate floor topping (approximately $\frac{1}{8}$in. thick) to the finished concrete. Epoxy coatings are difficult to apply, and therefore, an experienced flooring contractor should be obtained to install the finish. Another satisfactory "contaminated area" floor can be obtained by using Kalman, "absorption process" heavy duty cement finish floor. In all instances, whatever the floor finish may be, all pipes, ducts, drains, and other penetrations must be sealed.

Windows. In most climates, exterior wall windows can be a serious source of heat or cooling loss for the heating, ventilating, and air-conditioning system of the building. However, for architectural balance or to maintain the esthetic environment, it may be necessary to incorporate windows into the building design. Glass block masonry openings with small [12 in. × 18 in. (30.5 cm × 45.8 cm), 18 in. × 18 in. (45.8 cm × 45.8 cm), or 18 in. × 24 in. (45.8 cm × 61 cm)] double-pane insulating window units for view panels have been satisfactorily used in containment areas. The interior face of the glass block should be smooth. The exterior block should be flush with the wall to avoid providing roosting places for pigeons and other birds. All mortar used in glass block masonry construction, including the joints between glass blocks and metal surfaces, should be of a type to provide tight, nonshrinking, waterproof, corrosive-resistant joints. Whatever type of window is chosen it should not be capable of being opened in containment areas and in air-conditioned areas.

Lighting. Lighting in a microbiological containment facility should be designed to use fluorescent fixtures, except where waterproof, vaporproof, or explosion-proof fixtures are required. In the latter areas, suitable incandescent fixtures may be used. The minimum acceptable levels for illumination are:

Laboratories and similar work areas	75 ft-c
Work surfaces in laboratories	100 ft-c
Walk-in refrigerators and incubators	30 ft-c
Noncontaminated animal holding rooms	50 ft-c
Contaminated holding rooms	75 ft-c
Air locks	20 ft-c
Within cabinets and hoods	100 ft-c

Lighting fixtures can either be suspended from the ceiling or they can be recessed into the ceiling and sealed so that they can be serviced from above. The latter arrangement will mean that maintenance personnel need not enter the containment area to change lamps. However, this situation may require most of the floor space in the attic or service area above the containment area for the lighting fixtures. Obviously this arrangement is expensive and cannot be recommended except where cost is not important or where there are other overriding factors involved.

It is recommended that a programmed maintenance schedule be set up to service all fluorescent lamps, usually on a yearly basis. This program should negate the necessity for maintenance men to enter containment areas to change burned-out lamps on an individual basis. If suspended fluorescent fixtures are used, the covers should be of the hospital-approved, sloped type to prevent buildup of dust and contamination.

Standard daylight equivalent levels should be considered in windowless animal facilities.[69] This can be accomplished by providing a centrally controlled, timed, on-off lighting system. Alternatively, a simple timer can be installed in the light switch circuit to each animal room.

A system for emergency lighting must be provided. If there is an emergency electrical power system, minimal lighting can be connected to this system. If there is no emergency electrical power service available, then wet-battery, sealed-beam, flood-type emergency lamps should be provided in corridors, stairways, attics, basements, utility rooms, and other remote areas. These emergency lamps should be mounted on the wall and plugged into a nearby electrical outlet. If the power fails, these lamps will provide adequate illumination for a minimum of one hour.

Noise Control. Motors, pumps, compressors, air supply and exhaust systems, elevators, and individual pieces of laboratory or plant equipment such as vacuum pumps or centrifuges may create disturbing noises. Any mechanical equipment that creates excessive noise should be remotely located and, if need be, surrounded by noise-absorbing materials. Most noisy operations are found in the animal rooms and in the cage washing areas. Monkeys and dogs are also quite noisy. Other laboratory animals, while not creating noise, are greatly disturbed by excessive noises. Rubber-tired casters and rubber bumpers on carts, trucks, and racks will reduce noise. Construction materials should be selected to contain the noise generated in animal cage cleaning, storage, and preparation areas. The use of asbestos, cellulose, or plastic noise-absorptive tiles is not recommended in laboratory or animal room areas. Concrete walls have been found to be more effective than metal or plaster walls in containing sound.

Refrigerators. Most laboratory rooms will contain refrigerators of the free-standing, household type or of the below-the-benchtop type. Experience has shown that laboratory personnel store flammable solvents such as ether in these refrigerators. Since at the same time the refrigerator may contain infectious or toxic materials, explosions are a double hazard. For this reason, it is recommended that interior light switches and thermostatic controls be removed from all free-standing refrigerators and be mounted on the outside of the refrigerator.

Doors. Doors to laboratory rooms, animal rooms, storerooms, air locks, glassware washing rooms, cage washing rooms, and similar areas should be at least 42 in. wide. They should swing into the room and away from the corridor, except where doors are recessed and the door edge will not project into the corridor. All exit doors should swing in the direction of exit travel. Double doors designed to swing out in the direction of egress should be used for chemical laboratories where there is a risk of explosion and fire. Sliding doors are recommended where large equipment is handled or where large experimental animals such as sheep or cattle are being used.

Stainless steel armor plate (16 gauge type 430) should be placed to a height of four feet on all doors of laboratories and glassware washing rooms where carts and hand trucks are pushed through. It is recommended that automatic door closers (i.e., pneumatic or some other suitable type) be placed on all the doors within the building that need to remain closed to maintain correct ventilation and air balance.

Corridors. Personnel corridors should be a minimum of six feet wide, and ordinarily not wider than eight feet. Utility corridors should not be narrower than five feet and normally need not be wider than eight feet.

Floor Loading. While building codes vary in their requirements for floor loading, it is recommended that floors in most laboratory and plant areas be capable of withstanding a live loading of at least 80 lbs per sq ft.

Shop or Special Work Area. For most laboratory facilities, it is recommended that a small shop for minor repairs and/or alterations to equipment be provided. This shop might also include glass-blowing and repair capability and be located at some central point in the building. In such a small shop, minor repairs and alterations can be made by technicians or a full-time workman if tools are available. The concept of a local shop can reduce loss of time and inconvenience, since personnel and equipment do not need to leave the facility. However, a centralized, well staffed maintenance and repair facility will be required for major problems and alterations.

CHAPTER 4

Selected Mechanical Design Features

MECHANICAL CONSIDERATIONS

Heating, Ventilating, and Air Conditioning

General. The design of the entire air handling and treatment system for microbiological contamination control facilities must receive special attention. This system is one of the most critical in the facility, since it serves as part of the "secondary barrier," to control and maintain clean or sterile work areas for pharmaceutical product protection and for product sterility testing. The directed and specially treated air flows may also be used for operating rooms, patient isolators, tissue culture work, and to create clean biological conditions in spacecraft assembly facilities.

Functions of Air Handling Systems. Properly designed and balanced air handling systems and auxiliary equipment accomplish the special contamination control functions in containment facilities. Such systems can prevent the gradual buildup of concentrations of microorganisms in the air and can effectively isolate spaces within a building. Air handling in such facilities is ". . . one of the major problems of laboratory design perhaps because we try to justify its cost in terms of comfort factors. Comfort is a consideration but the function of 'air separation' is far more important . . . 'Air separation' means separation of laboratory spaces by providing a 100% outside air supply to each space with no recirculation of air. It can be achieved by providing a one-pass air handling system with separate supply and exhaust for each space. Absolute separation cannot be achieved except in sealed chambers; however, for practical purposes 'air separation' can be achieved. With a one-pass air handling system using the principle of air separation and auxiliary air handling equipment such as the fume hood and biological safety cabinet, the capability of . . . facilities can be expanded to include safe handling of viral, mycobacterial, and fungal isolations causing infectious diseases. . . ."[74]

Air handling problems must be discussed in the planning stages because air handling systems will markedly affect the layout of the facility.[137] Air handling problems can be specifically identified or categorized as accomplishing the following functions in a contamination control facility:

(1) Providing the normal functions of temperature control, humidity control, and air cleanliness.
(2) Preventing buildup of microorganisms in the air and removing or destroying airborne microorganisms.
(3) Isolating spaces within the facility and preventing cross-contamination between spaces.
(4) Preventing dissemination of potentially infectious microorganisms to surrounding community or area.
(5) Providing sterile work areas within the facility for special work situations or procedures.

The Use of 100 Percent Outside Air. Most laboratories requiring biological contamination control restrict the use of recirculated air and require 100% outside supply air. The primary justification for the use of 100% outside air is that it allows isolation of a room or suite after a biological accident and permits decontamination of the area without interruption of operations in adjacent areas. To facilitate the designer's understanding of this problem, the following is a brief description of procedures which might be employed for a decontamination process:

After an accident occurs in a laboratory, all personnel leave the room immediately. There should be a switch outside the room to close a damper or the louvers on the supply grill to the room. Discontinuing the supply air will cause the room to fall under a greater reduced air pressure than the surrounding laboratories. Formaldehyde solution (37%), or beta-propiolactone, in a portable mechanical vaporizer * can be placed in the room or chemical vapors can be introduced through ceiling or wall spray heads. The vaporizer is equipped with a timer and will turn itself off after treatment is completed. The reservoir of the vaporizer is filled sufficiently to give one cubic centimeter of formaldehyde per cubic foot of space, or one cc of beta-propiolactone for 5 cu ft of space. Even though the room is completely fogged with formaldehyde, little or no odor of formaldehyde will be detected in the hallway or adjacent rooms. If air were being recirculated, all personnel would have to leave the building, since formaldehyde is irritating at levels of 10 parts per million and beta-propiolactone is only slightly less irritating. If no shutoff damper on the supply duct to

*Challenger Model 5100 Vaporizer, Z&W Mfg., Co., Wickliffe, Ohio.

the room is provided, respiratory protection will be required for entering the room in order to cover the supply grill with a plastic sheet or heavy paper.

The use of 100% outside air is also justified to confine or dispel objectionable or toxic odors emanating from such activities or operations as animal rooms, cage and rack washing, anaerobic chambers, chemical reactions such as hydrogen sulfide, autoclaving of animal carcasses, etc.[46] Nonrecirculation of supply air also enhances and improves biological safety, by preventing recontamination of other areas in a facility with contaminants intentionally or accidentally released.

The previous paragraphs have delineated several of the most common reasons for the use of 100% outside air supply in a facility requiring biological contamination control. However, the designer should also consider, in cooperation with the user, the concept of recirculation of air providing such air is passed through ultrahigh-efficiency filters. This alternate approach requires a policy decision by the scientific staff, after a complete evaluation of projected operations, procedures, and agents to be used, an analysis of level of hazards, and a complete understanding of the abilities and limits of HEPA filters. It has recently been shown that only a very small percent penetration can be achieved through HEPA filters using a high loading of T_1 bacteriophage or microbial aerosol upstream of the filter.[56] In many facilities, these penetration levels may have a greater degree of safety than other portions of the barrier system, and therefore, recirculation through ultrahigh-efficiency filters could be the appropriate choice, possibly resulting in considerable economic savings as compared to a one-pass 100% outside air system. Most laminar air flow rooms are designed to employ the recirculated concept. These rooms usually contain personnel, and therefore, the minute amount of contamination which passes through the filter would be insignificant when compared to the loading from the human occupants. In addition, the cost of conditioning outside air *and* maintaining the extremely high air turnovers would be completely prohibitive. Therefore, a recirculated system is usually chosen in such instances.

Air Handling Systems. There are many components to any air handling system. Those that exert major influences on the operating characteristics of the air system are (1) the air tempering coils, (2) the fans, (3) the ducts, and (4) the auxiliary air handling equipment.

Air Tempering Coils: The coils found in air handling systems are: (1) Preheating Coils. These heat incoming outside air to prevent the freezing of the cooling coil and permit, in winter, the addition of moisture for humidity control. (2) Spray Cooling Coils. These control the moisture content of the incoming air and remove excess heat. (3) Reheat Coils.

These add heat to the air being supplied to the spaces for final temperature control.

As a rule, a filter is found before the first coil in order to remove particles which, in time, may interfere with the operation of the air handling equipment.

Fans: Included here are all fans or blowers supplying air to or exhausting air from the facility. These may be located at central points in certain zones, or in individual rooms. They may be supply fans or exhaust fans. Every fan or blower in the laboratory must be considered as part of the air handling system.

Ducts: Included are all ducts in the laboratory. If ducts carry air to the laboratory, they are supply ducts. If the carry air out of or away from the laboratory they are exhaust ducts. If they take air out of the laboratory and back to the supply fan room for recirculation of the air, they are return air ducts.

Auxiliary Air Handling Equipment: This part of the system includes devices such as fume hoods, biological safety cabinets, dehumidifiers, etc.

Each part of the air handling system must be considered in terms of its effect on the total design. In the final analysis, all components must function in a predetermined manner if the basic air handling objectives are to be achieved.

These basic components of air handling systems can be arranged in various ways to form a number of different systems, such as (1) single zone, (2) multi-zone, (3) fan coil, (4) reheat, and (5) dual duct systems. The last three systems are generally individual room control systems.

Single Zone System: The single zone air handling system illustrated in Figure 37 has one of each of the basic components controlled by a single

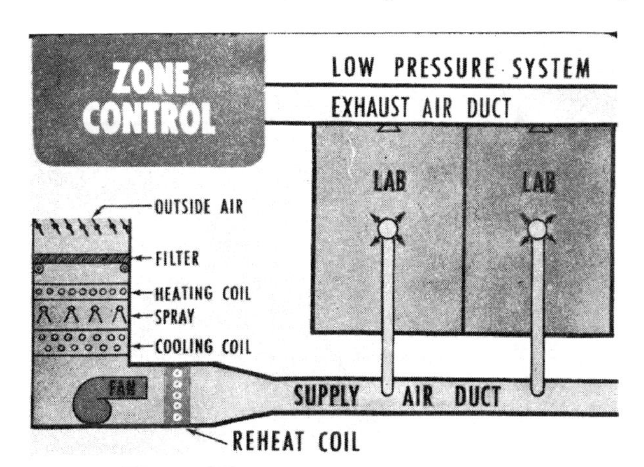

Figure 37. Single zone system.

thermostat located in a selected area. The temperature of the space in which the thermostat is located determines the temperature of the air supplied to all rooms served by the single zone system. This system can be designed to provide the required degree of air cleanliness and flow patterns using one-pass air; however, it cannot provide individual room control. Hence, it is usually not particularly suitable for multi-room facilities requiring microbial contamination control.

Multi-Zone System: The multi-zone system, not illustrated, has a single preheat coil, cooling coil, reheat coil, and supply fan, but has two or more thermostats. This system can be used where known differences in heat load occur. An example would be in a building where a number of rooms have glass in the exterior wall and others are completely surrounded by occupied spaces. The air to these two areas would be delivered in two separate ducts with each controlled by its individual thermostat. This system has the same basic characteristic as the single zone system. It can provide required air cleanliness but cannot provide individual room control.

Fan Coil System: The fan coil system is illustrated in Figure 38. This system consists of a number of small units located throughout the facility. While they share a common preheat coil, each unit has its own cooling coil, reheat coil, supply fan, and thermostat. With this system each area supplied by a fan coil unit is controlled independently of all other units.

The fan coil system is unique in that it provides for air separation of the space in which it is located from other spaces without the use of one-pass air. This is illustrated in Figure 39. This is possible since the fan coil unit is located in the building and air can be recirculated without affecting areas served by other units. Contamination buildup may be a problem in some

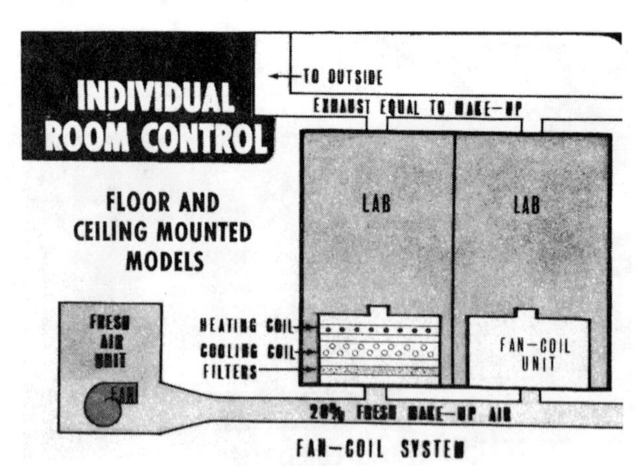

Figure 38. Fan coil system.

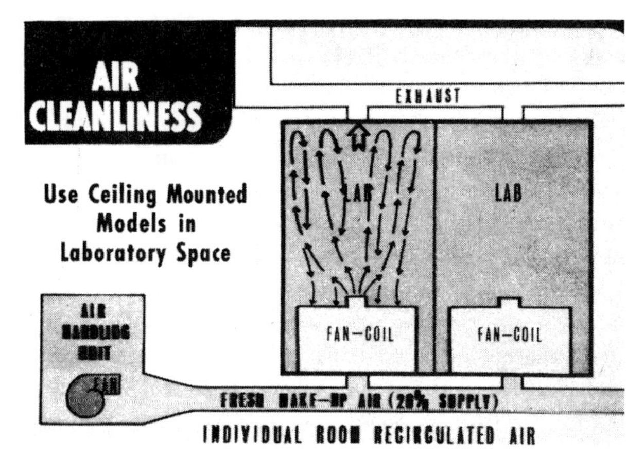

Figure 39. Fan coil system, recirculating laboratory air.

areas since it recirculates air in the space it serves. Fan coil systems meet most requirements for temperature and humidity control. While they may also be used for 100% outside air, under this condition, the units probably would have to be designed and fabricated for the specific job. Since standard units could not be used, another system may be a better solution.

Reheat System: The reheat system is shown in Figure 40. All basic components except the reheat coil and thermostat are located in a central point near the laboratory. The reheat coil and thermostat are located in the functional area of the facility, one for each room or space that is to be independently controlled. The principle of operation of the reheat system is that, at the central point, the air is treated to a predetermined tempera-

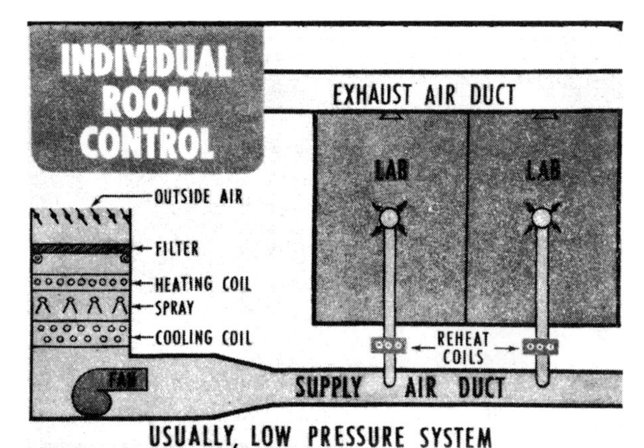

Figure 40. Reheat system.

ture, usually between 52° and 58°F. It is supplied to the facility at this temperature during all seasons of the year. In the room or space where the reheat coil is located, the air supplied by the supply fan room is heated so that the space can be maintained at the desired temperature. This system can provide the required air separation in a 100% outside air system and individual room control of temperature and humidity.

Dual Duct System: As illustrated in Figure 41, this system uses two ducts, one carrying hot air, the other cold air. At the room, the air from the two ducts enters a mixing box controlled by the room thermostat. The thermostat controls the proportions of hot and cold air brought into the mixing chamber so that air leaving the chamber maintains the space at the desired temperature. A constant volume box is provided so that the volume of air being supplied to the room remains constant. While this system differs greatly in actual design from the reheat system, it has the same basic capabilities for air separation, air cleanliness, flow patterns, and individual room control of temperature and humidity. An additional advantage is that it does not get off balance and cause reversed air flows.

Building Supply and Exhaust Systems. While it is the supply system that often receives the most attention during the design of the air handling system for a containment facility, the exhaust must not be overlooked. Adequate design of the exhaust is essential for air separation of space; and where 100% outside or one-pass air is used, the exhaust system must handle the same quantity of air as the supply system.

The exhaust system can be classified as either central, zone, or local, depending upon the areas handled. A central exhaust is one in which the

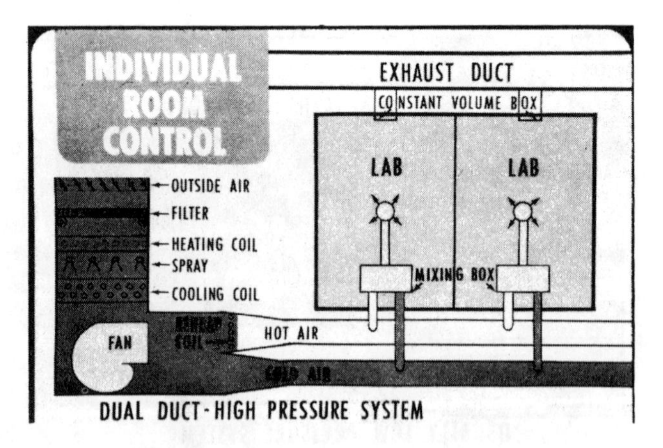

Figure 41. Dual duct system.

return air from all laboratories is collected by a central exhaust fan system and discharged through one common exhaust stack. A zone exhaust system is one which handles a segment or one area of the building system. Local exhausts are usually required for fume hoods, and chemical storage cabinets, and microbiological safety cabinets, and may exhaust directly to the roof with no connection to the main exhaust system.

The central or zone exhaust should be used wherever possible since this provides maximum flexibility for the location of equipment such as microbiological safety cabinets and fume hoods. With local exhaust, when it becomes necessary to add or relocate this equipment, a new duct to the roof is required. If central exhaust is provided, it is a simple matter to install the equipment and connect to a nearby central exhaust duct.

Finally, it should be recognized that the location of the supply intake and exhaust grills is a critical facet in the system design. This will be especially true with a 100% outside air system. When the central or building exhaust stack discharges effluent air at or near the building supply fan, it may be taken back into the laboratory; hence, a one-pass system, which theoretically separates spaces, may become a recirculating system without the required separation. Occasionally, the use of a high-speed discharge fan may aid in the separation process. As a general rule, the building supply and exhaust should be located as far apart as possible, horizontally and vertically. Selection of proper stack height and discharge velocity are equally critical to avoid reentry or short-circuiting. Further, the exhaust grill should be located downwind from the supply intake. There are other equally important considerations with respect to planning building intakes and exhaust, such as ratio of potential fume hood site to laboratory space, treatment of supply and exhaust ductwork etc., that have been previously discussed.[118,135]

Treatment of Exhaust Air. Air exhausted from those microbiological contamination control facilities where infectious, potentially infectious, or toxic materials are handled should be treated in some manner to remove or inactivate the material under investigation. This treatment process, another of the secondary barriers, can be accomplished by several methods; however, the most common approaches that have been utilized are filtration and/or incineration. The designer and the user (scientist) must carefully evaluate various factors such as the agents to be studied, the techniques to be employed in their study, location of the facility, etc., and together select the appropriate treatment methods.

Filters with varying levels of efficiency are commercially available and have proven effective for filtration of microbial particles.[30] It must be recognized that the exhaust air from general laboratory space will, in general, contain a very small amount of infectious or toxic material, since the majority of operations involving such material will be conducted in appropriate containment equipment such as microbiological safety cabinets. Therefore, it may prove practical to select high-efficiency filters [30] for exhaust from the general laboratory area rather than ultrahigh-efficiency filters. If an appropriate design has been selected for the exhaust stack [page 117], the efficiency of the filter (approximately 99%) and the dilution factors should provide adequate protection.

When filtration of air exhausted from microbiological contamination control equipment, such as safety cabinets, etc., is employed, it may be decided to increase the efficiency of the filters, since the level of risk has also increased and greater concentrations of infectious material may be present in such exhaust air. Commercial filters, known as ultrahigh-efficiency or HEPA filters, have been microbiologically evaluated for penetration of submicron T, bacteriophage, which has number mean diameter (NMD) of .1 micron.[56] Ultrahigh-efficiency filters utilized at rated capacities are designed for a face velocity of approximately 5 ft/min. The results of the tests that were conducted indicated that the percent penetration varies from 0.002% for one manufacturer to 0.00002% for another manufacturer. Such filters, therefore, provide excellent protection against viral and other microbial aerosols.

In those instances where intentional aerosols of highly pathogenic organisms are artificially created in specialized chambers such as the Henderson apparatus, it may be desirable to employ incineration of the exhaust air. Since the incinerator could fail to operate and therefore could pass microorganisms without adequate treatment, an in-line filter is usually employed as a fail-safe device. An appropriate length of pipe for adequate "dwell-time" of the microorganisms in the critical area of the incinerator is determined by the volume of exhaust air per unit time.

The seals and joints of exhaust ducts for air that is treated to remove potential or infectious material must be designed and fabricated to prevent accidental escape of hazardous material. The ducts can be constructed of stainless steel or of galvanized sheet metal. The joints should be taped and epoxy coated or welded and sealed pressure-tight as determined by a soap bubble test at 4 in. of water gauge. If the exhaust ducts are located outside the facility, soldered or welded joints must be employed. When filters are used, provision must be made for microbiological decontamination of the filters prior to changing of the filter by maintenance personnel. If a dual plenum at the filtration point is provided, gasketed, gastight dampers will allow diversion of the exhaust air through a new filter, and decontamination *in situ* without interruption of the air balance for the facility. A satisfactory method is to employ a steam/formaldehyde mixture, at the rate of one cubic centimeter of formaldehyde per cubic foot of space. If air flow must be maintained through the filter during decontamination, an additional cubic centimeter of formaldehyde must be added for each cubic foot of air passing through the plenum. Aeration of the plenum for several hours will be required before maintenance personnel, without a ventilated suit, can enter the plenum for filter changing.

Other methods of air treatment for microbiological contamination control, such as electronic precipitation or UV sterilization, have been employed. However, in most instances such systems require extensive, expensive maintenance, and have not been demonstrated as effective as filtration or incineration. In some cases, it may be necessary to include additional types of air treatment for removal of toxic or noxious fumes to supplement the microbiological treatment process.

The most obvious weak link in any air filtration system is the possibility of leaks, holes, etc., that could allow infectious material to bypass the filter. Proper gasketing and sealing of filter units to the filter frames, together with periodic microbiological and/or physical testing of all filters used in the contamination control system will eliminate this concern. Testing of the effectiveness of incinerators on a periodic schedule is also required. Testing must be done in such a manner that a break of sterility is not involved.[30,55]

Ventilation Requirements of Ventilated Safety Cabinets. The partial barrier open panel bacteriological cabinet should be ventilated at a rate of at least 75 linear feet per minute through the opening. The exhaust air should be drawn through a high-efficiency filter and then through a blower mounted outside the laboratory room, preferably in the attic. The blower should discharge the air into the central exhaust system upstream from the central exhaust filters. If the room is small or if several partial barrier cabinets are manifolded together after their filters, the exhaust blower should operate continuously to maintain a constant air flow from the laboratory room. When the cabinet is used with the front closed, greatly reducing the cabinet air flow, then a bypass opening in the front of the cabinet exhaust duct, downstream from the cabinet filters, should open and keep the blower at a constant flow rate.

Exhaust ducts from all bacteriological ventilated cabinets (partial or absolute barrier) should be maintained at a reduced air pressure relative to the laboratory to prevent backflow from the duct into the laboratory. Such a backflow could wash an exhaust filter and bring potentially infectious material into the laboratory.

Absolute barrier, gastight cabinets are checked for leaks by pressurizing to 2 in. water gauge (w.g.) with Freon gas and checking all surfaces and joints with a G.E. Type 2, halogen leak detector. When these cabinets are used, they should be maintained under $\frac{3}{4}$ in. w.g. negative (less than laboratory atmosphere) pressure. Inlet and outlet filters from these cabinets are of the ultrahigh-efficiency type. The minimum ventilation rate should be 10 air changes per hour (4 cfm per standard $30 \times 40 \times 36$-in. module) or sufficient to limit the temperature rise due to internal heat load to 10°F, whichever is greater. Exhaust air from the absolute barrier cabinet is carried in gastight, welded steel ducts or galvanized ducts with taped epoxy-coated joints.

System Reliability. All components of the mechanical systems for microbial contamination control must have sufficient redundancy to maintain their function, even with mechanical failure of major items of equipment. Such requirements may seem expensive and unnecessary until one calculates the actual investment in animals, biological cultures, long-term research materials, or specialized equipment, all of which are protected and maintained by these mechanical systems. In some cases, the object behind the microbial contamination control barrier is irreplaceable, as might be the case with research animals inoculated with rare vaccines or potential tumor material, or in the case of a component for an extraterrestrial exploration vehicle, whose performance and reliability can be insured only if specific environmental factors remain constant.

There are several primary utilities whose prolonged loss could produce a major crisis. The provision of uninterrupted electrical service should be considered of prime importance. In almost all regions, it is possible to provide electrical service to a facility from at least two independent power stations. In the event of a localized failure, this may prevent equipment outage. However, in most instances, it will be necessary also to provide generators for auxiliary power supply in case of a power failure. Such generators should be connected so as to automatically energize circuits handling the essential equipment in critical areas such as animal facilities, patient isolation facilities, or other key areas.

In almost any region, the loss of heating or cooling capability during critical seasons could be as catastrophic as a total power loss. It may be difficult to provide complete standby capability for such a situation, but consideration of such problems early in the design process may allow at least a partial solution. When boilers are required for heating or other critical services, it may be feasible to design some excess capacity into the system and then install two or more boilers rather than one large unit. In the event of a failure of one boiler, the remaining one(s) could be used to provide steam for critical areas or functions of the facility. It is also possible to rent or lease large boilers to provide additional capacity during emergency situations. Likewise, provision of cooling redundancy may be accomplished by splitting the required load between two or more chillers and handling only the critical cooling requirement in the event of an emergency. Portable cooling units are commercially available and have been used by commercial airlines for several years. On occasion, used units which become available might provide an inexpensive standby cooling source for large institutions.

An additional consideration for complete system reliability is an adequate supply of critical spare parts. Some components of the microbial contamination control system may be rare, complicated, and difficult to obtain on short notice. The facility designer should alert the user to such situations so that critical unit components can be stocked for emergency situations.

Piping

Water Service. All water (except drinking water) supplied to microbiological contamination control facilities for infectious material study should pass through a break tank within the facility. This device will prevent accidental contamination of the central water supply by a reverse flow of water. In some cases, a backflow preventer may be substituted for the break tank. If a backflow preventer valve is used, it should be of the type that has two spring-loaded vertical check valves and one spring-loaded, diaphragm differential pressure relief valve, as approved by the State of California Health Department. A backflow preventer valve must be tested and serviced frequently by an expert.[21]

A safety shower with a spring valve should be installed in each laboratory that contains a chemical fume hood. Other safety showers should be strategically located throughout the laboratory area. The safety shower should be equipped with a chain pull to facilitate its use by a person who is temporarily blinded because of a laboratory accident. The water supply for the safety shower, as well as for water stills, personnel showers, and wash basins, should come from a line connected to the water main before the break tank. All drinking fountains (foot-operated valves to prevent unnecessary contact with water supply) should be similarly connected, unless supplied on a separate line.[13]

There should be two hot water systems in an infectious disease laboratory building: one system to serve the laboratory area and the other system to serve all clean areas, including personnel showers and lavatories. The system serving the laboratory area should be supplied with water downstream from the break tank. The clean area system will obtain water directly from the main, upstream of the break tank.[51]

Since internal microbial contamination of piping and water storage tanks can occur, distilled and/or deionized water sources should be close to the point of use. Central sources of distilled and/or deionized water involve lengthy piping and water storage in contaminated or potentially contaminated tanks. The contamination factors and the resulting metabolites in the water may outweigh the simplicity of centralized water systems. Therefore, careful consideration should be given to generation of such water at the point of use. If a central system is required, careful selection of the generation equipment, the storage system, and the piping system materials is required. An excellent review has recently been made of distilled water distribution systems.[117]

Another type of water service that is required in laboratories is the fire sprinkler system, when allowable by the local codes. Fire sprinkler systems should be installed in storerooms, change rooms, offices, contaminated and noncontaminated receiving rooms, chemical laboratories, and attics. Sprinklers should not be installed in infectious disease laboratories where there is a risk of increasing the spread of infectious material, nor should sprinklers be installed in rooms containing electronic equipment or other expensive equipment that would be damaged by water. In these instances, some type of gaseous extinguishing system should be employed.

Compressed Air. Experience and knowledge of compressed air systems has demonstrated that there is little or no danger of reverse flow or cross-contamination. Therefore, compressed air can be supplied by a single system to both contaminated and noncontaminated areas. The compressor should be of the oil-free (carbon or Teflon* ring) type, and equipped with an after-cooler, an air receiver, and a pressure-regulating valve downstream of the receiver. Compressed air lines and outlets, when required, are provided in laboratories above the laboratory benches and in the glassware washing room. Pressure and flow requirements are usually 5 cfm at 40 psig at each station.

Vacuum. Separate vacuum systems for contaminated and noncontaminated areas should be used if possible. If this is not possible, the two legs of the central system should be separated by a high-efficiency or ultrahigh-efficiency filter to prevent biological cross-contamination. Vacuum service outlets are also provided on bench tops in laboratories and in all glassware washing rooms. The vacuum system should be capable of pulling at least 27 in. of mercury. The vacuum receiving tank should have a drain connec-

* Teflon ®—Dupont Chemical Co., Wilmington, Delaware.

tion for draining accumulated liquid into the contaminated drain system. There should be an in-line filter before the vacuum pump.

Gas. Propane or natural gas is usually required on all laboratory bench tops in biological contamination control facilities for study of infectious material. A single system can serve both contaminated and noncontaminated areas. The partial barrier, open front cabinet can be serviced with gas, but special provisions must be made for an absolute barrier gas-tight cabinet system if it is to be serviced with gas. In the latter case, the gas line must be equipped with an automatic or spring-loaded "dead man" shutoff valve, since such equipment may at times be pressurized, and it is at least remotely conceivable that reverse flow could occur. Whenever possible, operation in absolute barrier cabinets should be conducted with portable gas canisters, alcohol lamps, or electric heaters to minimize contamination control problems and explosion hazards.

Drains. Normally, there are three categories of liquid wastes from a biological facility: storm sewage, sanitary sewage, and contaminated sewage. The storm sewage consists of normal rainwater runoff. The sanitary sewage from noncontaminated areas of an infectious disease research facility is treated in a conventional sewage treatment plant. In those facilities where highly infectious materials are studied, the contaminated sewage is collected from the various drains in the contaminated areas of the building and is treated to inactivate the biological materials before discharge into the conventional sanitary sewage line. (For a more detailed discussion of this topic see pages 136–144.)

The drains on the noncontaminated side of the change room should be connected to the sanitary sewer. The showers and all drains after the showers in the contaminated change rooms, laboratories, etc., should run into the sewage treatment facility.

All laboratory rooms, corridors, animal rooms, and any other area that will be washed and treated with liquid disinfectants should have a floor drain. All traps in the contaminated area should be extra deep to prevent negative pressure from pulling the traps. It is recommended that all floor drains be provided with a piped priming water supply to each trap. Floor drains should be supplied with a screw-type plug which may be inserted when the drain is not in service for long periods of time. If floor drains are installed in animal rooms, they should be of the non-clog, bucket type. Schedule 40, welded, wrought iron pipe should be used for all drains from equipment that is sealed to the room, such as a biological safety cabinet. All drains from sinks, sterilizers, and chemical fume hoods, open to the room (except floor drains), should consist of screwed fittings to a point past the trap; thereafter, the lines should consist of Schedule 40, welded, wrought iron pipe.

Vents. All vent lines should be fabricated from Schedule 40, welded, wrought iron pipe. There should be no cross-contamination of vent lines from different floors, and each floor should be vented separately. Vents from the same floor should be manifolded in the attic and run through the roof as a single vent, with filter adapter, block valves, and steam connections. Upon completion of all vent piping systems in a new building, a smoke test should be made of both the contaminated and noncontaminated systems to insure against cross-contamination in these systems.

Mechanical Services Distribution

Introduction. The essential mechanical services, including electrical service, for any facility requiring biological contamination control may amount to more than half the total project cost. In addition, maintenance costs for such facilities are relatively high and have been increasing each year. Minor alterations will always be required as programs change, and occasionally major modifications to a facility may eventually be required. Therefore, the distribution of mechanical services for these facilities must be carefully planned to insure: (1) moderate initial procurement and installation costs; (2) ease of maintenance; (3) flexibiilty in design to allow rapid, inexpensive, and easy installation of minor alterations; and (4) ability to make major utility changes without requiring major structural modifications.

The first objective in the design of the mechanical distribution system is to satisfy the functional needs, both immediate and future. However, it may be difficult for the architect to determine such factors as electrical requirements for each room, the compressed air and vacuum requirements, the natural or propane gas requirements, and hot and cold water, as well as distilled water, requirements. The problem is further complicated because the time lag between the planning of a facility and its occupancy may be several years, during which time programs, procedures, and personnel may change drastically. With the increasing use of automation and of electronic equipment in biomedical research, electric power requirements and equipment usage will continue to increase, while conventional bench space will decrease. The mechanical distribution system must be designed to meet these changing needs, and should permit some isolated major increase in demand without affecting the remainder of the system.

For example, many new facilities do not provide sufficient air supply and exhaust for additional fume hoods, bacteriological safety cabinets, or other ventilated enclosures. The designers have provided adequate air for all ventilated enclosures anticipated during the design phase, but if additional cabinets and hoods are needed, the air required for ventilating these enclosures is not available without destroying the air balance in the facility.

Mechanical Services Layouts. *General.* Figure 42 shows a typical biological laboratory consisting of two side wall benches, a peninsular bench with sink, and equipment space for such items as a refrigerator, freezer, incubator, and centrifuge. Two windows are shown in the outside wall. Windows not only limit the bench and equipment layout, but also limit the type of mechanical distribution system that can be utilized.

Figure 43 shows this same laboratory with the mechanical services normally required in such a laboratory. Shown are electrical outlets, cold water, hot water, natural or bottled gas, compressed air, vacuum, room supply air, and room exhaust air. The sanitary waste system with vents is not shown. Since these mechanical services may reach the laboratory in a variety of ways, five mechanical services distribution systems are illustrated and discussed in this report.

Ceiling-Floor System. Figure 44 illustrates one of the most common mechanical services distribution systems. In this system the services are

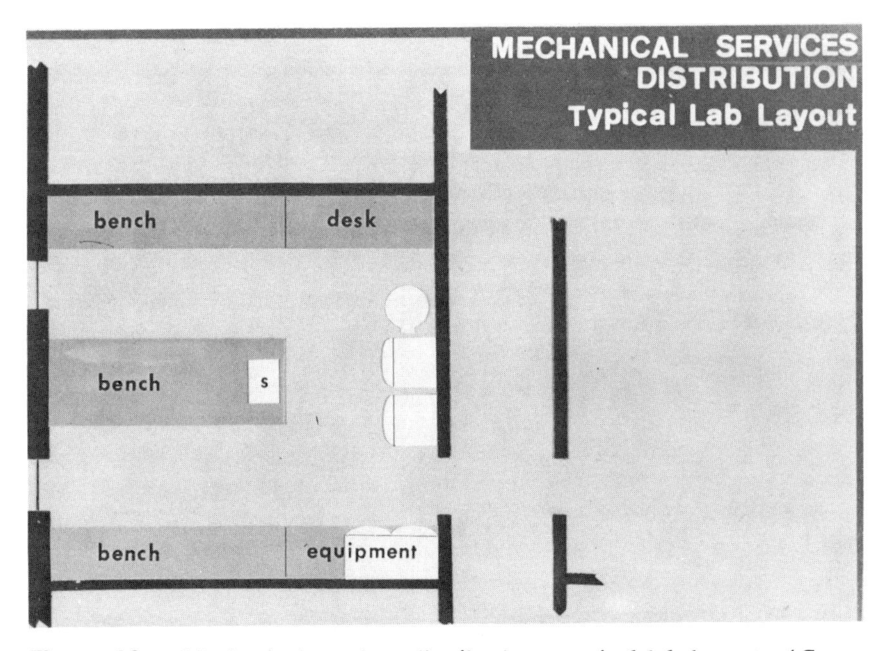

Figure 42. Mechanical services distribution—typical lab layout. (*Courtesy U.S. Public Health Service.*)

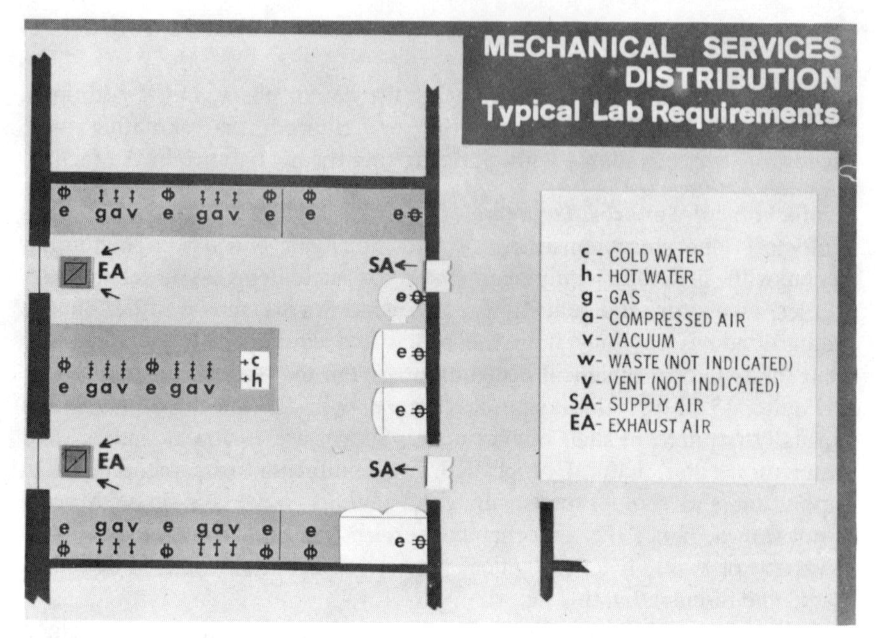

Figure 43. Mechanical services distribution—typical lab requirements. (*Courtesy U.S. Public Health Service.*)

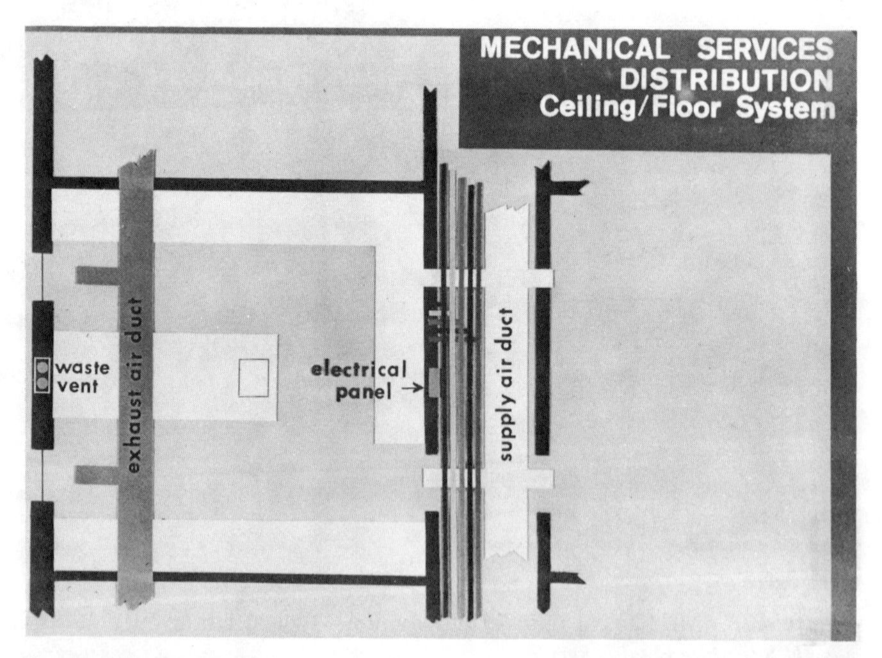

Figure 44. Mechanical services distribution—ceiling-floor system. (*Courtesy U.S. Public Health Service.*)

shown running horizontally in the corridor, either below the corridor floor or above the ceiling in the corridor. In this illustration, the exhaust duct is shown in the laboratory since there is not sufficient space in the corridor. Figure 45 shows the ceiling-floor system in cross section. In this illustration, the services, except ventilation, pass from the corridor through a utility space and through the floor up to the work sites. The services can, of course, be brought down from the utility space above the ceiling as shown in Figure 46. However, this would require exposing the pipes in the laboratory or providing some alternate type of concealment. The laboratory arrangement would determine whether it would be better to supply utilities downward from above or upward from below. The ceiling-floor, utility space system has the dual advantage of having the lowest initial cost and requiring the least floor space. However, room layout changes are difficult, major changes are costly, maintenance is complicated, and floor penetrations are numerous. If further construction economies are necessary, cost reductions can be made by omitting the ceilings and running all pipes exposed without sacrificing function. The appearance may be objectionable to some, but in the biological contamination control facility false ceilings

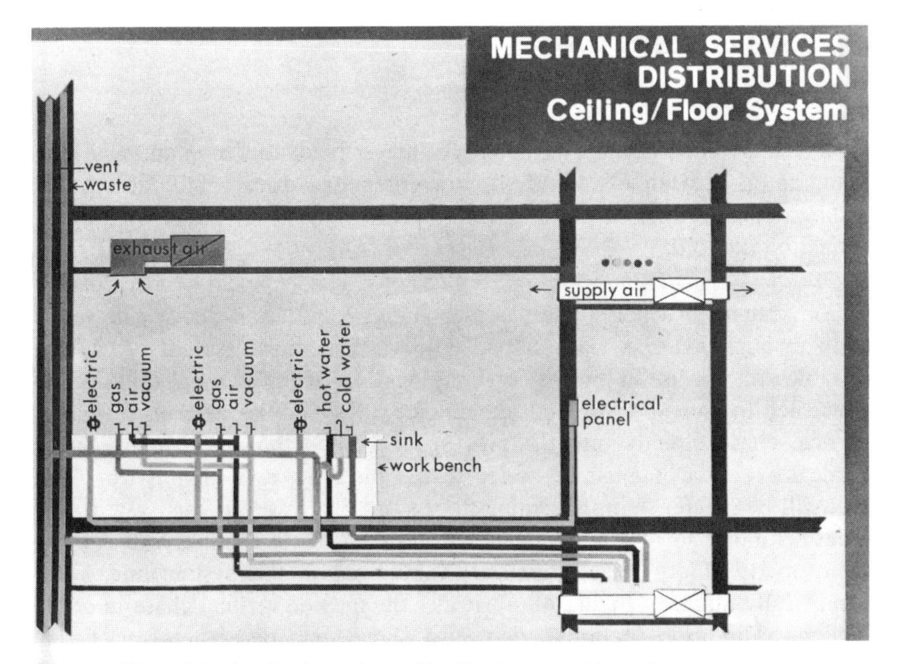

Figure 45. Mechanical services distribution—ceiling-floor system. (*Courtesy U.S. Public Health Service.*)

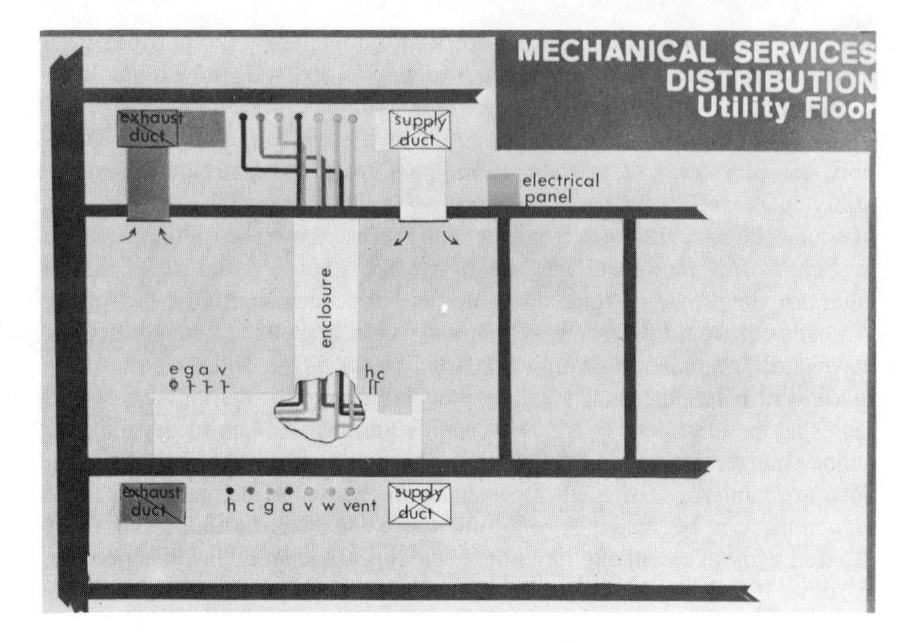

Figure 46. Mechanical services distribution—utility floor system. (*Courtesy U.S. Public Health Service.*)

should be avoided because these spaces are difficult to decontaminate and maintain in a clean condition. In the infectious disease laboratory, the ceilings should either be solid and hermetically sealed or the service pipes should be exposed.

Outside Vertical Chase. Figure 47 shows a mechanical services distribution system in which all services, including ventilation ducts, run vertically through exterior wall shafts. The vertical chase system is usually suitable only for use in multi-story buildings. In this system, all utilities are concealed to give an improved appearance. The utilities are run from the vertical chase directly into the pipe space behind the workbench. Since vertical appendages must be added to the building, it is obvious that the cost will be greater than the ceiling-floor system. However, this system has great flexibility so that minor changes and changes in room layout can be easily made. Floor penetrations are minimized in this system and walls can be fully utilized. Figure 48 illustrates the outside vertical chase in cross section. This cross-sectional view also shows two disadvantages of the vertical chase. First, it is difficult to service and maintain and the section of workbench adjacent to the chase may have to be removed to get to the

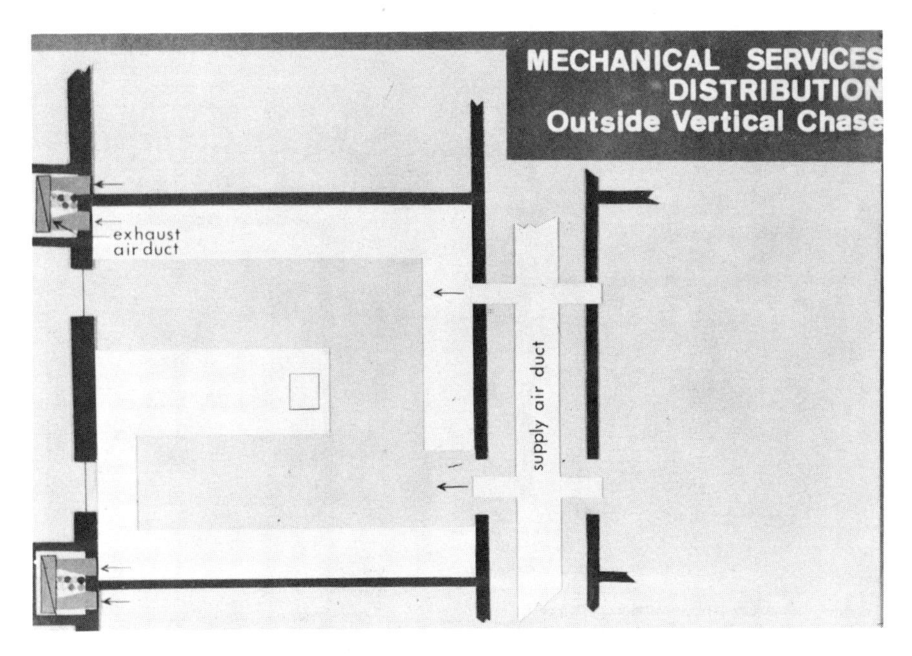

Figure 47. Mechanical services distribution—outside vertical chase. (*Courtesy U.S. Public Health Service.*)

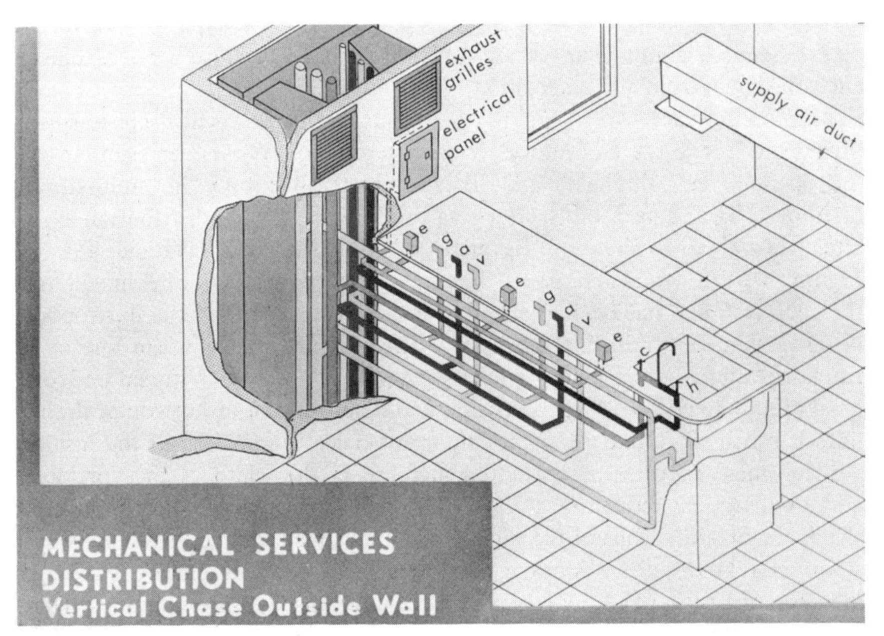

Figure 48. Mechanical services distribution—outside vertical chase. (*Courtesy U.S. Public Health Service.*)

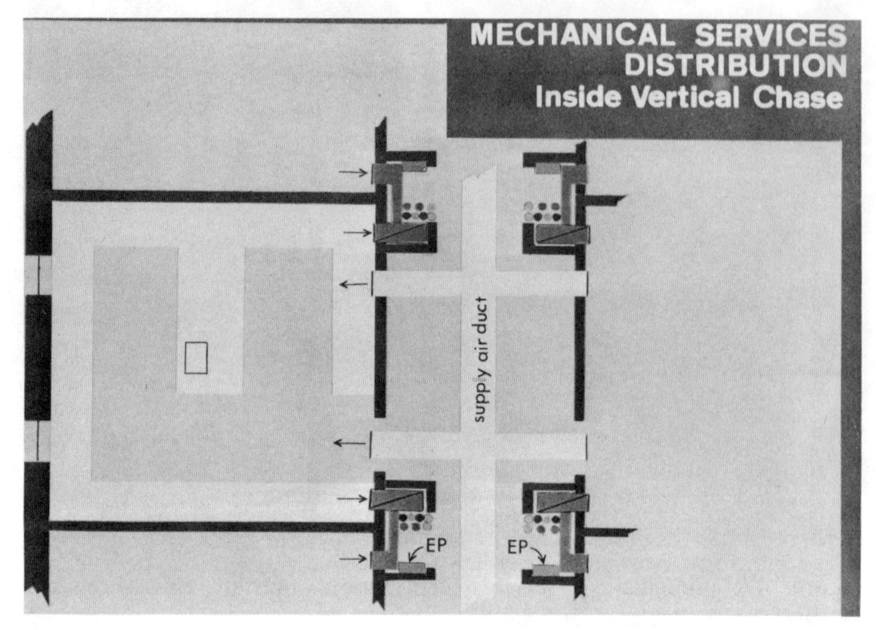

Figure 49. Mechanical services distribution—inside vertical chase. (*Courtesy U.S. Public Health Service.*)

pipes. Second, major changes are difficult to make because services must enter the room from a fixed position.

Inside Vertical Chase. Figure 49 depicts an inside vertical chase supplying services to the laboratory. This system is, of course, very similar to the outside vertical chase, and likewise is feasible only in multi-story buildings. The same room layout was used in the service distribution systems shown in Figures 42 through 48. In the inside vertical chase, Figure 49, the peninsular bench has been moved from its attachment to the back wall and has been attached to a side wall to shorten the horizontal distance that the drain line must run. The distance that a waste line can run horizontally before being vented is governed by various national and/or local plumbing codes. Also, corrosion and deterioration of horizontal drain lines is much more likely to occur than in vertical lines. Use of the inside vertical chase requires a substantial increase in the width of the corridor and a consequent reduction in the depth of the laboratory rooms on either side if the exterior walls of the facility are unchanged. Some of the hallway space can be used for storage, but the inside vertical chase definitely increases the ratio of utility space to laboratory space in the building. The inside vertical chase has the same advantage as the outside vertical chase.

All utilities are concealed and minor changes and changes in room layout can be made easily. Floor penetrations are minimal and walls can be utilized fully. The disadvantages of the outside vertical chase in being rigid and inflexible, as far as major building modifications are concerned, also applies to the inside vertical chase. The difficulty in maintenance of the outside chase can be overcome in the inside vertical chase as shown in Figure 50. Doors opening into the corridor can be mounted on the inside vertical chase to simplify maintenance, repairs, or the addition of new services.

Horizontal Gallery. A mechanical services distribution arrangement in which a utility gallery is placed between the outside wall and the laboratory is shown in Figure 51. The outside service gallery uses extra floor space and is therefore more costly. It decreases the laboratory area ratio as compared to the gross area. This system can be used in one- or two-story laboratory buildings, but it is most suitable and economical for use in multi-story buildings. Although the outside utility corridor has a high initial cost, it has several real advantages. Maintenance and repair are excellent as can be seen in Figure 52. All pipes, ductwork, and electrical lines and panels are readily available for maintenance. Maintenance per-

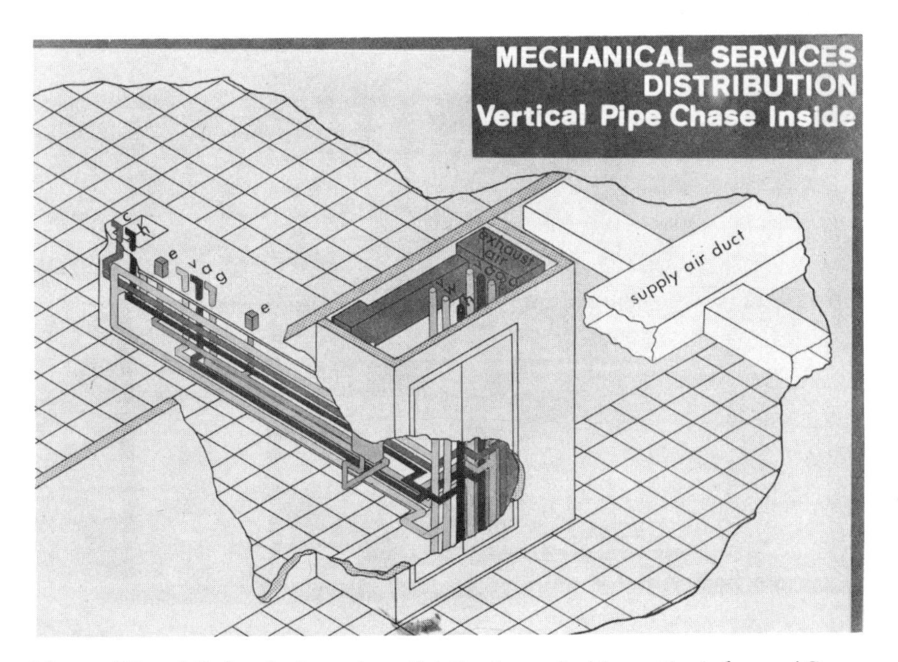

Figure 50. Mechanical services distribution—inside vertical chase. (*Courtesy U.S. Public Health Service.*)

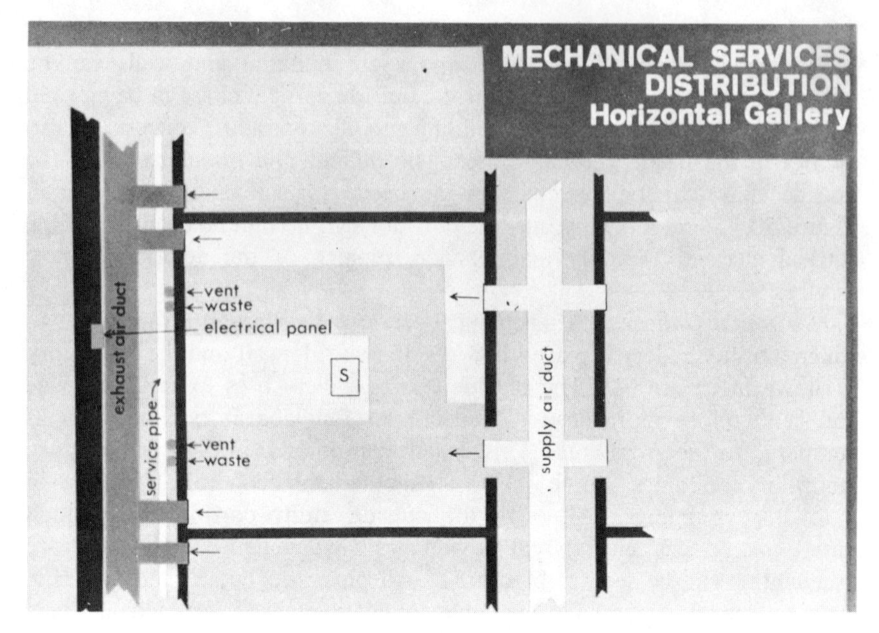

Figure 51. Mechanical services distribution—horizontal gallery chase. (*Courtesy U.S. Public Health Service.*)

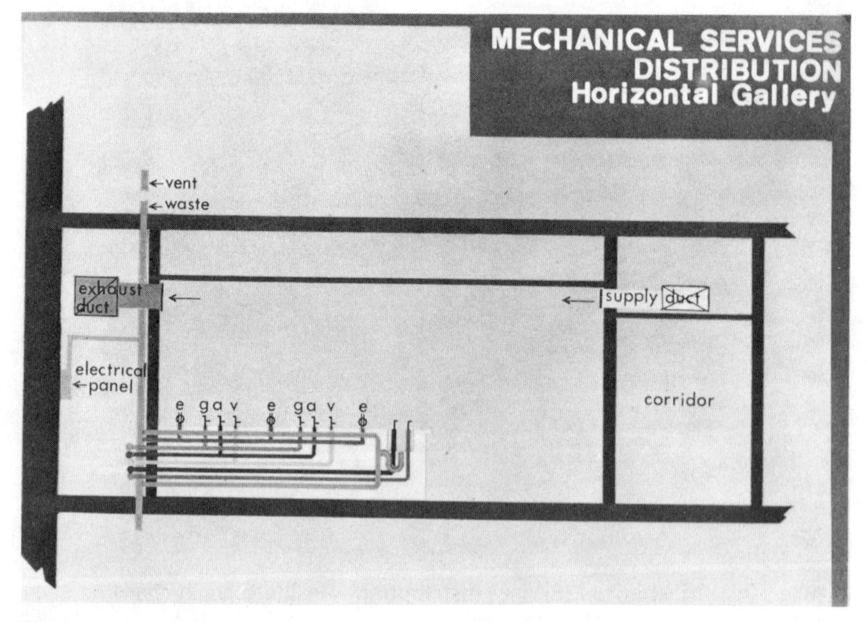

Figure 52. Mechanical services distribution—horizontal gallery chase. (*Courtesy U.S. Public Health Service.*)

sonnel do not need to enter the laboratory area to accomplish their service or repair work. In the infectious disease laboratory or in facilities requiring a high level of clealiness, this concept is especially desirable because it reduces the possibility of unnecessary exposure. The horizontal corridor has great flexibility, and minor modifications in the building are made easily. Also, major building modifications can be made with few changes to the distribution system. A second building can be added in parallel, with the existing horizontal services gallery serving both buildings.

Utility Floor. The utility floor, mechanical services distribution system is illustrated in Figure 53. This system provides concealed utilities with a common ductwork and drainage system in a manner that permits almost unlimited flexibility; however, it does so at a high initial cost and a low net to gross area efficiency. All service mains and ductwork are brought to the various utility floors by means of centrally located shafts, from which distribution is made laterally to any point of the floor above or ceiling below. The laboratory room is sandwiched between the two utility floors. Figure 53 shows the workbench located in the center of the room, which demonstrates the flexibility of the utility floor system. Figure 45 illustrates this concept with a cross section of a laboratory between two utility floors. The utility floor system is not a good selection for other than multi-story

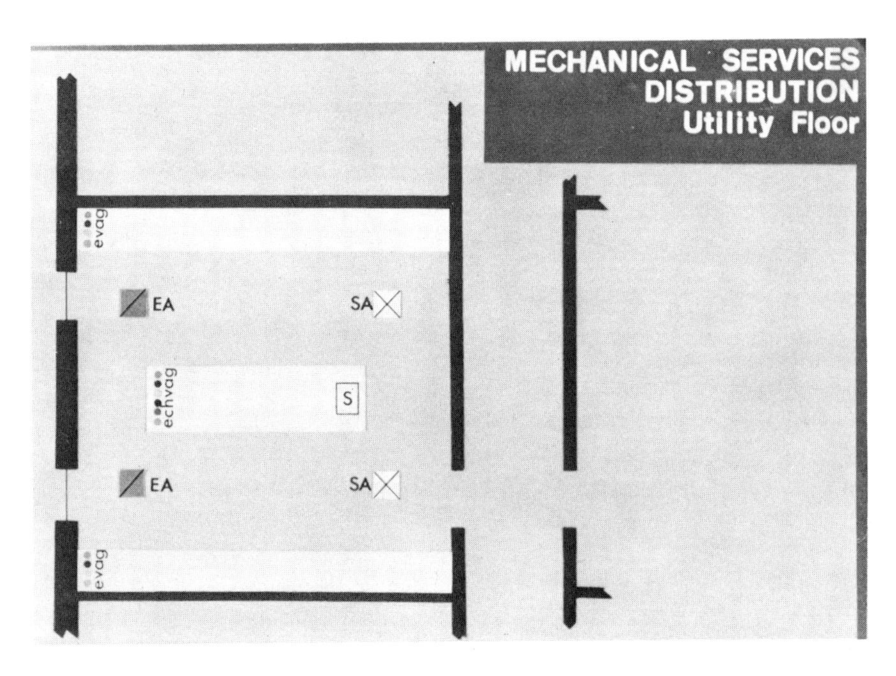

Figure 53. Mechanical services distribution—utility floor system. (*Courtesy U.S. Public Health Service.*)

buildings. It permits either minor or major modifications to the building with little or no modification in the mechanical services distribution system.

Other Distribution Systems. There are several other methods of service distribution, such as a pipe trench to a center island or services bracketed along a back wall. However, the five distribution systems discussed in this report are the systems most often used in biological laboratories.

Systems Analysis. Figure 54 establishes six basic objectives of a distribution system and gives a rating for each of the five systems in meeting these basic objectives. First, design is listed under all five systems since any system can satisfy the functional need, if the need can, in fact, be established. Second, the five systems vary greatly in their ability to meet the objective of being changed easily for minor alterations or change in room layout. The ceiling-floor system is difficult to change to allow even minor modifications. The outside or inside vertical chase systems are good in meeting minor changes or changes in room layout. The horizontal gallery is excellent in allowing minor modifications, because all the services are available along the entire back wall of the laboratory. The only problem is that some bench work must be removed where the workbench is against the back wall. The utility floor system is superior in allowing

MECHANICAL SERVICES DISTRIBUTION – Systems Analysis

OBJECTIVE	CEILING/ FLOOR	OUTSIDE VERTICAL CHASE	INSIDE VERTICAL CHASE	HORIZONTAL GALLERY	UTILITY FLOOR
SATISFY FUNCTIONAL NEED	DESIGN	DESIGN	DESIGN	DESIGN	DESIGN
FACILITATE CHANGE IN LAYOUT	POOR	GOOD	GOOD	EXCELLENT	OUTSTANDING
PERMIT CHANGE IN UTILITY DEMAND	DESIGN	DESIGN	DESIGN	DESIGN	DESIGN
MAJOR UTILITY CHANGE - MINOR STRUCTURAL CHANGE	POOR	FAIR	FAIR	EXCELLENT	OUTSTANDING
SIMPLIFY MAINTENANCE	POOR	FAIR	GOOD	EXCELLENT	OUTSTANDING
MINIMIZE FLOOR PENETRATION	POOR	GOOD	GOOD	EXCELLENT	GOOD

WHERE DESIGN IS INDICATED, THE OBJECTIVE IS ACCOMPLISHED BY THE DESIGN OF THE SYSTEM AND IS INDEPENDENT OF THE DISTRIBUTION SYSTEM USED.

Figure 54. Mechanical services distribution—systems analysis. (*Courtesy U.S. Public Health Service.*)

minor modifications to be made in a laboratory. Because services can be brought in from any point in the ceiling or floor of the laboratory, almost any layout arrangement can be accomplished readily. Third, a local change in utilities demand can be accomplished only if the system will permit it. The architectural engineer must have had sufficient information or foresight to plan for such a change in demand in his original design. If plans for a change in utility demand have been made, then any system will permit the local increase in demand without adversely affecting the remainder of the system. Fourth, the five distribution systems vary greatly in allowing major modification in the building structure. The ceiling-floor system is poor because ceilings will have to be lowered and floors penetrated. Both vertical chase systems are fair in allowing major modifications. The horizontal gallery with adequate working room is flexible and, as with minor modifications, it is outstanding in allowing major modifications. Fifth, the ease of carrying out maintenance on the various systems ranges from poor to outstanding. If utilities are covered in the utility floor system, the maintenance will be difficult. If utilities are exposed, then a higher maintenance rating would have to be assigned to the ceiling-floor system. The inside vertical chase can be rated higher than the outside vertical chase, if the former is equipped with a door so that workmen have easy access to the services. The horizontal gallery allows complete access to all services and can be rated as excellent in the maintenance category. The utility floor system provides easy access to all of the services with ample room to carry out the work. Sixth, floor penetrations are numerous in the ceiling-floor system and any modifications are likely to result in new holes in the floor. The outside and inside vertical chases are good because most lines will run above the floor to the chase. The horizontal gallery is excellent with regard to floor penetrations, because only rarely will a pipe need to pass through the floor. The utility floor system can be rated as good because only drain lines will need to penetrate the floor.

System Economics. Figure 55 rates the five mechanical services distribution systems with regard to initial costs and space requirements. Basically, there are two types of costs involved. One is the initial cost of the pipe, valves, ductwork, and other physical items of equipment, plus the cost of installing this equipment. These materials and labor costs are essentially the same for all systems, except that considerably longer runs are required in some systems to bring the services to the workbench. The second cost of the services is in the space required for the system. Each square foot of space of the structure has a cost and the total space available is limited. The ceiling-floor system requires only short runs of piping, and no extra space for services is required. On the other hand, in the utility

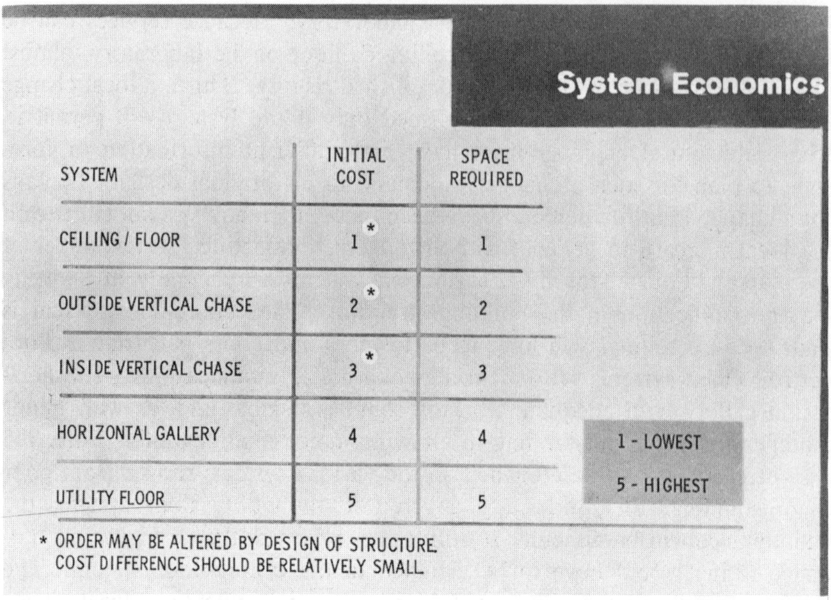

System Economics

SYSTEM	INITIAL COST	SPACE REQUIRED
CEILING / FLOOR	1 *	1
OUTSIDE VERTICAL CHASE	2 *	2
INSIDE VERTICAL CHASE	3 *	3
HORIZONTAL GALLERY	4	4
UTILITY FLOOR	5	5

* ORDER MAY BE ALTERED BY DESIGN OF STRUCTURE.
COST DIFFERENCE SHOULD BE RELATIVELY SMALL.

1 - LOWEST
5 - HIGHEST

Figure 55. Mechanical services distribution—system economics. (*Courtesy U.S. Public Health Service.*

floor system, the services will have to run greater distances and large amounts of floor space for carrying the utilities are required. The ceiling-floor system is the least costly initially and requires the least space. The utility floor carries the greatest initial cost and requires, by far, the greatest amount of space. The other systems fall somewhere between these two systems in initial cost and space requirements. The initial cost may vary somewhat among some of the systems due to the design features of the installation. Outside vertical chases could be quite expensive to construct, whereas a horizontal gallery, because of the unfinished nature of the space, may be relatively inexpensive to construct.

SPECIALIZED FEATURES

Liquid Waste Treatment Systems

General Rationale for Sewage Treatment. The liquid effluents from an infectious disease research laboratory or a vaccine production plant will, on occasion, contain some of the infectious materials under study or production. These viable microorganisms will enter the building sewage system as the result of an accident or spill, through the failure of a piece of equipment or the carelessness of a worker. Ordinarily, a small amount of infectious material entering a sanitary sewage system will not create a significant infectious hazard since the material will be sufficiently diluted and will probably be inactivated as it passes through the sewage treatment plant and is chlorinated before being discharged to a stream. However, if the research laboratory is producing liter or gallon amounts of material for enzyme or other studies, as often occurs, then treatment of liquid effluents from the laboratory should be considered. Vaccine production laboratories would fall into this category and should have the capability for sterilization of sewage. In addition, some city or municipal building codes require treatment of microbiologically contaminated effluents.

The nature of the infectious agent will influence the decision on sewage sterilization. Exotic agents not endemic in the area must be handled with greater care. For example, foot and mouth disease virus and rinderpest virus cannot be brought into the continental United States for research or for any other reason. Agents of great epidemic potential, such as those causing plague and smallpox, must be handled with great care. From a public or community relations standpoint, the cost of a treatment system may be justified if the facility director can state that all avenues of escape of organisms have been eliminated. The microbiological laboratory handling infectious materials is in a vulnerable position from the public relations point of view, and it is likely to be blamed for various illnesses in man or domestic animals that occur in the community. Whether proven or not, such publicity can be damaging. Therefore, the entire barrier system in such a facility *must* be reviewed from a biological safety standpoint, not simply the system for treatment of sewage effluent. Such factors as effluent air treatment, sterilization (or disinfection) of equipment and materials leaving the facility, and special clothing and other personnel protective devices must also be accomplished in order to insure overall consistency in microbiological safety.

The type of central sewage treatment employed will influence the decision as to whether local laboratory treatment should be carried out. Some workers have recovered infectious agents such as *S. typhosa* and polio viruses from the effluent of treatment plants using a primary treatment system.[75,79] Others have shown that Coxsackie viruses, *M. tuberculosis,* and *S. typhosa* could be recovered from the effluent of treatment plants using primary treatment methods.[64] These workers also showed that Coxsackie viruses and tubercle bacilli could be recovered from sewage treatment plants using secondary treatment when final chlorination was inadequate. Tubercle bacilli were recovered 100 feet downstream from the treatment plant outfall. These various studies give ample evidence of the fact that pathogenic agents will pass through either a primary or secondary treatment plant to contaminate streams if the final chlorination is not heavy and if exposure to the chlorine is not sufficient.

Greater emphasis is being placed on various controls to prevent stream and river pollution. Some cities, such as New Orleans, Louisiana, are including in their city codes certain restrictions on infectious materials being discharged into the city sanitary sewage system. These types of restrictions will undoubtedly increase in the future.

Methods of Sewage Treatment. *Ionizing Radiation.* Cobalt 60 or other gamma-ray emitters could be used; however, the problem of handling and shielding strong radioactive materials are so great as to make this method impractical at the present time.

Chemical Treatment. Problems of mixing and holding chemicals for a period of time are considerable, and the necessity of inactivating the chemical after the sewage has been sterilized would make this treatment method expensive.

Ozone. Ozone could be used to sterilize sewage; however, this gas is difficult to handle, and operating costs for such a system would be high.

Other Methods. Ultrasonic systems have not been developed to the point where they could be used effectively. Ultraviolet sterilization of sewage is effective when the suspended solids have been reduced by digestion and/or filtration. All UV liquid sterilizing units should be equipped with UV intensity meters and controls to insure that the unit will cut off if the UV intensity drops below the effective level.

Heat. Heat is by far the most reliable, effective, and most economical method available for sterilizing sewage. When treatment of effluent sewage is required, it is the system of choice.

Types of Sewage Sterilization Systems. *Batch Sterilization Tank.* The batch sterilization tank (see Figure 56) consists of a steel vessel for collecting liquids that is equipped with valves that are closed before the contents are heated. After a holding period, the tank is allowed to cool or is cooled physically and the contents are drained into the sanitary sewer. The batch tanks are usually installed in series so that one tank can be filling while the other is undergoing sterilization. Steam passing through a heater, such as a jet heater, is usually used as the source of heat. The jet heater also provides adequate mixing of the tank contents. During the filling cycle, displaced air must be exhausted through a microbial filter or air incinerator. Instrumentation for this system consists of a liquid level gauge and a thermistor or thermocouple.

Treatment by this system usually consists of running the tank up to 30 psig steam pressure at the appropriate temperature and holding for

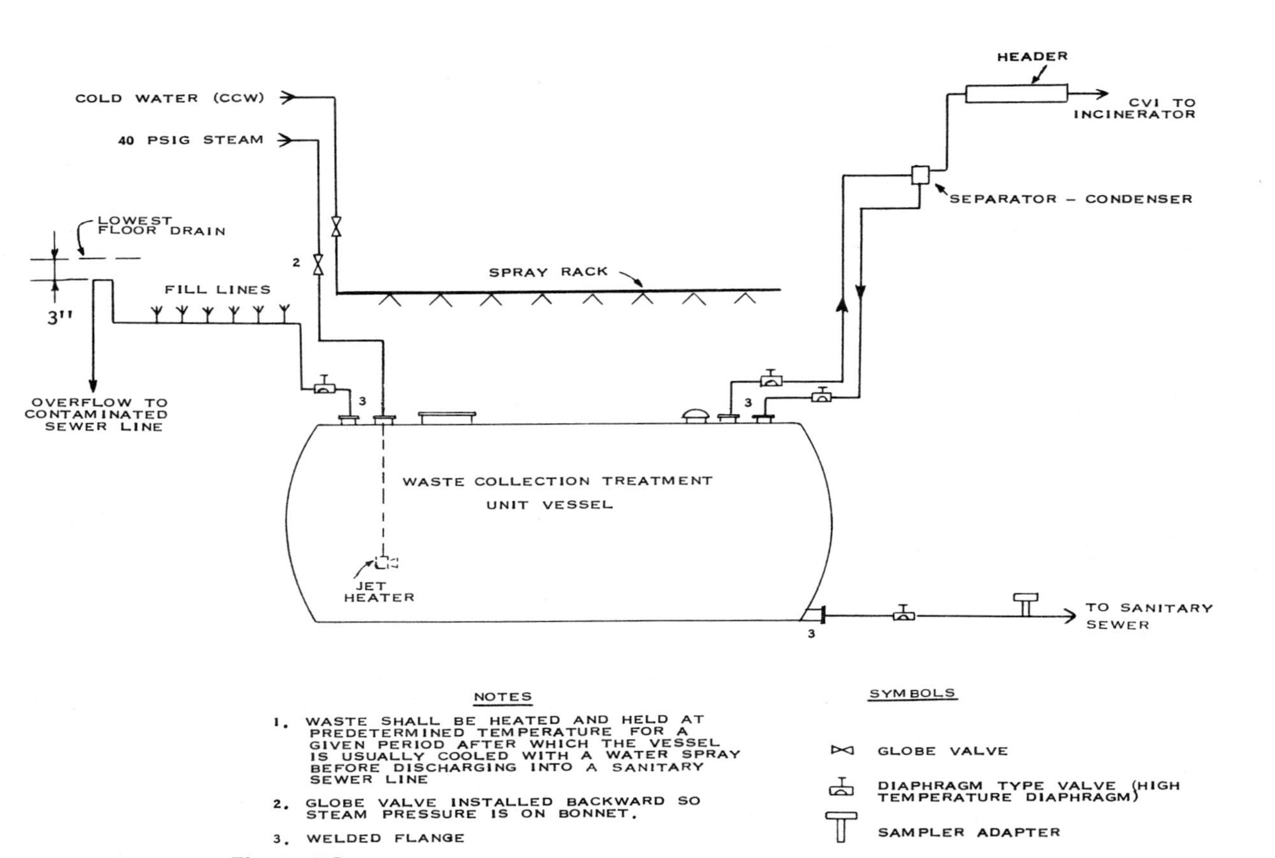

Figure 56. Waste collection treatment system—batch operation.

30 minutes. After heating, the contents of the tank must be cooled below the boiling point to prevent the liquid from flashing to steam as it is released. A cooling system such as dilution with normal sanitary sewage or a jacketed tank with chilled water can be installed. The batch system of waste treatment is rather expensive and wasteful, because no recovery of heat is normally attempted. Any size vessel can be used, but it is recommended that the vessel be large enough to retain a one-day flow (maximum) from the laboratory.

Continuous, High-Temperature Pasteurization System. The high-temperature pasteurization system is relatively inexpensive to procure and install, convenient to use because it requires minimum maintenance, and more economical to operate than the batch-type system. However, the system heats to only 195°F, as compared to 274°F (at 30 psi) for the batch system. Whereas the high-temperature pasteurization system will kill vegetable bacteria, viruses, and rickettsia, it will not inactivate bacterial spores or other heat-stable biological products.

The high-temperature pastuerization system (see Figure 57) consists of a holding and treatment tank, high and low liquid level probes, a steam heater, and a retention tube drain line. As liquid drains into the treatment tank, the high-level probe is activated, the drain valve is opened, and the contents of the tank drain out until the low-level probe is exposed and the drain valve is closed. An automatic temperature controller activates the jet heater to maintain the temperature of the liquid in the tank between 195° and 200°F. The retention tube lying in the bottom of the tank is designed to give a minimum of one minute exposure time to liquids that might be entering the treatment tank when the tank is in the draining part of its cycle. If local codes or restrictions prohibit discharge of treated sewage at high temperatures into the sanitary sewer, it may be cooled with a heat exchanger or cold water jacketed pipe, or diluted with cold water or sanitary sewage.

The waste collection treatment tank should be sized to retain a four-hour flow from the laboratory. Installation of dual systems should be considered so that one system can be cleaned and repaired while the second is in operation.

Constant Flow, Heat Exchanger Sterilization System. The constant flow, heat exchanger sewage sterilization system is the most professional, practical, and efficient method for sterilizing sewage (see Figure 58). This system is ideally suited to short-term heat treatment of liquids. The contaminated, low-temperature sewage picks up heat from the high-temperature sterile sewage so that the heat that must be added to the system is con-

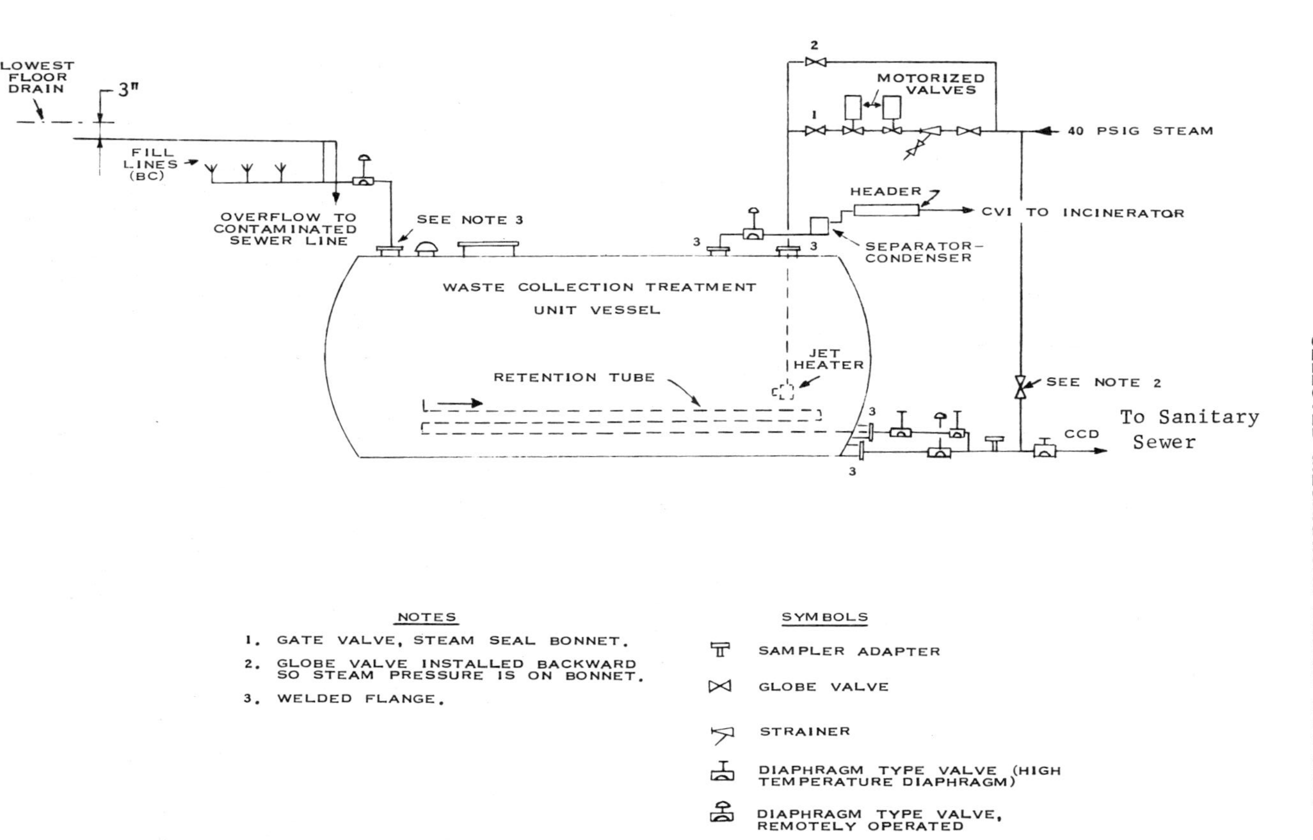

Figure 57. Waste collection treatment system—continuous operation.

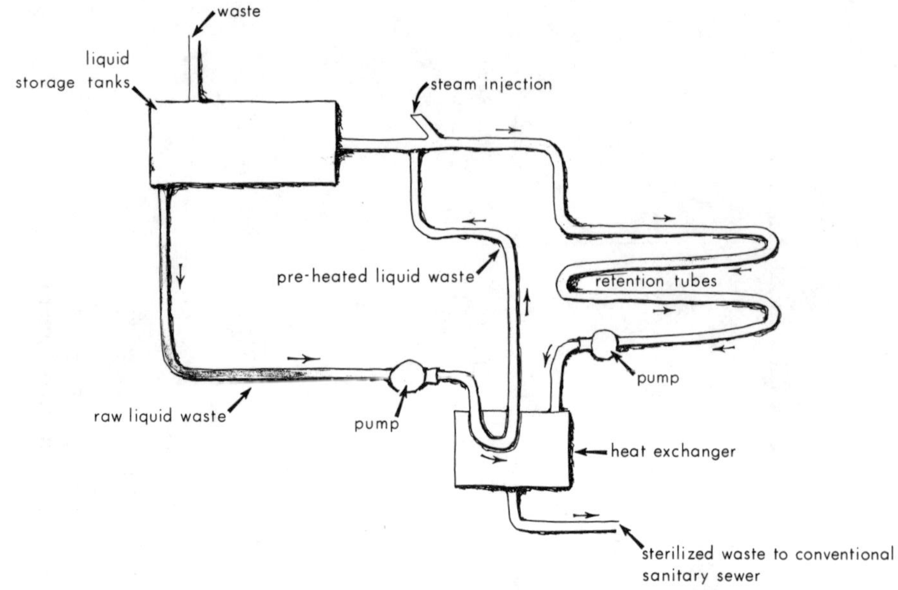

Figure 58. Constant flow, heat exchanger sewage sterilization system.

served. The sterile sewage is cooled to a point where it can be discharged into the sanitary sewer system.

Liquids from the laboratory areas usually are drained by gravity into a holding tank. A comminuter may be placed in the drain line before the line empties into the holding tank or tanks. When sufficient volume has accumulated, the system is started by operating the two circulating pumps. Steam is added through the steam injector, and the treated sewage is passed back into the holding tank instead of being discharged into the sanitary sewer until full operating temperatures have been reached. When operating temperatures have been reached, the treated sewage will be shunted into the sanitary sewer. A second pump, located behind the retention tubes but in front of the second pass through the heat exchangers, is essential. This pump raises the pressure of the treated sewage so that if a leak occurs between the tube and shell side of the heat exchanger, sterile sewage will leak into contaminated sewage and not vice versa. A rubber diaphragm type of sampling adapter should be placed on the discharge line, and samples should be taken at scheduled intervals to check on the sterility of the treated sewage.

Desired temperatures throughout the system, as well as retention time and percent of heat recovery, can be varied by the design and materials of the system. Heat exchangers can be obtained in almost any size, capacity, and shape, and can be fabricated from a variety of materials.

The heat exchanger system of sterilizing sewage is the method of choice when large volumes of liquids are to be sterilized from several laboratory buildings or a large facility. Procurement and installation costs are high, but operating costs are low. This system should be considered for use even when volume flows are quite small, since it has distinct advantages over the tank sterilization systems.

Sewage Sterilization Equipment Rooms. *Location.* Contaminated sewage is usually drained to the holding tanks by gravity; therefore, the sewage sterilization rooms will usually be located in basements or subbasements. Sewage holding tanks should be located in watertight pits, as shown in Figure 59, that are large enough to hold the entire contents should the tank rupture or a valve leak.

Drain System for the Sterilization Room. A sump with both a steam ejector pump and an electric pump, for liquids that might enter from an accident, leak, or overflow back into a second sewage holding tank, should be provided.

Figure 59. Batch treatment tank with pit. (*Courtesy U.S. Public Health Service.*)

Location and Testing of the Holding or Treatment Tank. The holding or treatment tank must be anchored to the concrete saddles or to the floor so that the tank will not float if the tank room fills with liquids. A treatment tank must be reinforced to take full vacuum if it is connected to a steam line, and must be hydrostatically tested at one-and-one-half times its maximum operating pressure. This test should be carried out when the tank is first installed and at periodic intervals during its use. The tank and lines should be tested for tightness before a new system is put into use. This test can be accomplished by using a chlorofluorohydrocarbon (Freon) and a G.E. Type 2, Halogen Leak Detector.

Miscellaneous. The equipment room should be ventilated for personnel comfort and should be maintained under a reduced air pressure. Air exhausted from the area should be passed through a separate filtration system or through the central air filtration system. If feasible, control panels, main valves, etc., should be located in an anteroom or should be remotely located to allow shutdown of the system without entry ino the space in the event of an accident. Personnel entering the room should wear laboratory-type clothing and obey all safety rules and regulations applicable to the laboratory. Gas masks and/or ventilated personnel suits should be available in the event of rupture or leak in the system or some other accident resulting in microbiological contamination.

Disposal of Contaminated Flammable Solvents

Introduction. The safe and efficient disposal of flammable waste solvents is often a problem of a research laboratory. When the quantities are large and the waste is contaminated with infectious agents, the problem becomes more difficult. Several of the research laboratories, such as those at Fort Detrick, Frederick, Maryland, utilize a procedure involving the solvent extraction of water from microbiological suspensions. The waste generated by the extraction process contains equal parts of acetone and ether and varying proportions of water. It is also highly contaminated with infectious agents. The Fort Detrick disposal system is described below.

System Concept. The system for disposal incorporates primarily commercially available equipment. An existing large-volume air incinerator, operating on a continuous basis, is used to sterilize exhaust air from contaminated equipment in several buildings. Adjacent to the incinerator is a small building containing an autoclave that is used intermittently for other purposes. The flow diagram in Figure 60 shows the disposal system integrated with existing equipment. The container of waste solvent is placed in an autoclave and connected to a welded line that runs over to a two-fluid

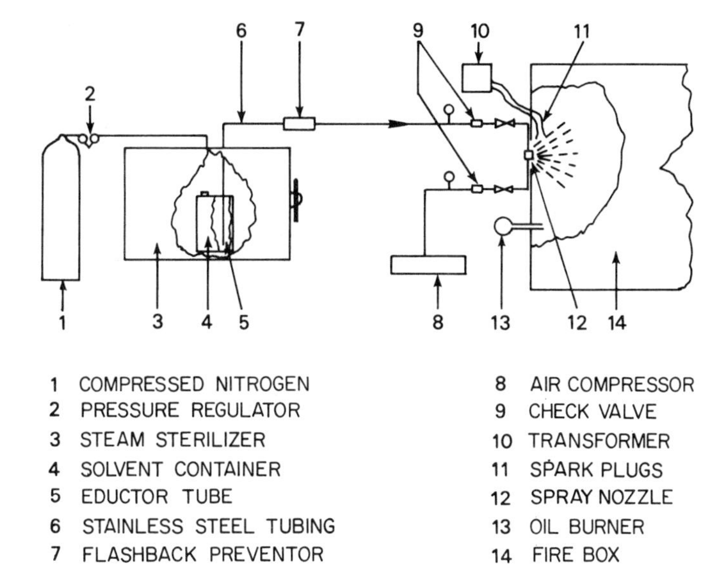

Figure 60. Disposal system for flammable solvents containing infectious disease agents.

spray nozzle inserted through the front surface of the incinerator firebox. This line has a flash arrestor, pressure gauge, and check valve.

Operation. Nitrogen pressure is applied to the autoclave through a regulating valve, forcing the solvent through the line and spray nozzle. Nebulizing air is supplied to the spray nozzle by an air compressor. To insure burning, a 10,000-volt ignition transformer feeds power to two industrial-type spark plugs modified and positioned so as to provide a continuous ignition spark. Optimum feed rate of solvent to the nozzle is about 10 gallons an hour. When the container is emptied of solvent, the system is sterilized by pressurizing the autoclave with steam and bleeding the steam through the nozzle for a period of about one hour.

Some types and sizes of containers that have proven practical are illustrated in Figure 61. The two on the left are commercially available pressure vessels that contain an eductor tube and are fitted with valves on the inlet and outlet. Future systems will be standardized on this type for quantities of 10 to 15 gallons of contaminated solvents. If greater quantities, such as 50-gal batches, must be treated at one time, it would be necessary to develop larger containers.

Figure 61. Solvent disposal containers. (*U.S. Army photograph.*)

The autoclave, container, and nitrogen components of the system, with a container connected to the solvent line, are illustrated in Figure 62. The spray nozzle and spark plug transformer, with a sight tube in the center for observing the flame, are shown in Figure 63.

Conclusions. It may seem redundant to utilize both the spark plugs as an internal source of ignition and the incinerator as an outside source. However, the composition of the solvent-water mixtures being disposed of can vary over a wide range of ratios. In addition, due to the low solubility of ether in water, a two-phase mixture is present in the container, water-acetone-ether in the bottom layer and acetone-ether on the top. As the water-containing layer is drawn off, incomplete combustion occurs at the sparkgap, and the incinerator provides the necessary heat to sterilize this noncombustible liquid. When most of the water has been eliminated and the liquid becomes more pure in solvents, the sparkgap provides a source of instant ignition to prevent a possible buildup of an explosive atmosphere in the incinerator.

The system described has proven sound and has performed satisfactorily. When such a system is included in a new facility, a small autoclave could be installed, rather than the large one shown in Figure 62, to dispose of small quantities of solvent.

Figure 62. Autoclave, container, and nitrogen components. (*U.S. Army photograph.*)

Figure 63. Spray nozzle with transformer (A) and sight tube (B). (*U.S. Army photograph.*)

Design Checklist

Throughout the preliminary and final stages of the design of a new research building, the planning group will be called upon to make decisions on a great variety of subjects. Experience has shown that a checklist can be very valuable to the planning group. The checklist in Appendix II is intended to assist in eliminating the possibility of overlooking any of the more important details connected with the design of the building.

Many of the headings are suggestive in that they will apply in various ways and in various parts of the building. These general headings should be explored fully according to the particular needs of the proposed facility under consideration.

CHAPTER 5

Facility Acceptance Procedures

A final acceptance procedure, regardless of complexity, cannot in itself produce a satisfactory facility for microbial contamination control. This can be achieved only through a proper and adequate design. In-process inspection and cleaning requirements are also important means of insuring the ultimate satisfactory performance of a facility. However, final acceptance procedures are necessary to determine if design requirements have actually been incorporated into the completed facility. While in-process inspection and cleaning requirements are usually of an engineering nature and pertain to removal of particulate matter, they also will accomplish microbiological objectives, if properly selected and implemented.

The guidelines in this chapter represent current methods and requirements necessary to produce a bioclean facility which, in turn, will simplify maintenance and support requirements. Any in-process inspection, cleaning requirement, or final acceptance test must be detailed and set forth as a contractual requirement or specification. This should be true whether a facility is constructed by contract or by an "in-house" task force. Since most microbial contamination control facilities are constructed by outside contractors, the following requirements are contract-oriented, and represent the state of the art.

MATERIAL CERTIFICATION

All materials, parts, subsystems, and systems should be submitted by the contractor for approval and should be accompanied by a certificate of compliance with design drawings and specifications. This certificate provides the necessary assurance that the items going into the construction are, in fact, acceptable to the original design. Changes or substitutions are generally allowed provided sufficient data are supplied for evaluation of equality.

IN-PROCESS INSPECTION

Masonry

This is an area of prime importance. Surfaces requiring smooth finishes by steel troweling, i.e. walls prepared for painting, floors, etc., must be free of troweling buildups, loose particles, or other residue. Such surfaces, improperly prepared, mean high filter maintenance and difficult acceptance resolution. Moisture content, if excessive, likewise can destroy an otherwise good paint job and be an endless source of problems, with particle contamination due to sealing and the decontamination difficulties created by a porous surface. If the wall, floor, or ceiling is to be a part of the secondary barrier system, all cracks or joints that are filled with an impervious sealant to prevent ingress/egress of microorganisms must be carefully inspected.

Metals

All metal surfaces should be nonshedding. Anodizing treatments or prime and seal treatments which have been specified must be inspected for compliance during construction since it may be impossible to correct such items after construction has been completed.

Sealing

Taping or caulking of joints, crevices, and leaks should not be allowed. Some improved mastic compounds are available, but the approved method is by compression of a gasketing material. This is particularly true of ultra-high efficiency filter frames. While many companies use compressible gaskets, heli-arc welding is preferred. Flush-mounted wall outlets and other penetrations through a microbiological barrier should be internally sealed and isolated from the hollow wall environment by silicon rubber mastic.

Plumbing and Ductwork

Both plumbing and ductwork are excellent sources of particulate matter during and after construction. Rust, scale, and oxides formed by electrolytic action are difficult to control. Sealing by a prime and seal method and judicious use of gaskets, as well as care in mating dissimilar metals, can aid significantly in reduction of such particulate contamination. The use of epoxy paints has minimized this problem for practically all materials. Even in the return side of an air handling system, shedding can be a problem by accelerating the loading time of prefilters, thereby increasing maintenance costs.

Routine Cleaning

There is considerable benefit in carrying out a routine cleaning program during construction. Areas that should be included and specified contractually are:

(1) Interior of all stud partitions.
(2) Walls, before application of gypsum board.
(3) Areas above ceilings.
(4) Interior and exterior surfaces of all ductwork.
(5) Interior and exterior surfaces of all electrical fixtures, races, ducts, and boxes.
(6) Collecting surfaces, such as window sills, door frames, steps, and floors.

Cleaning should be done weekly, and if a central vacuum system is specified, this could be installed early in construction and be used for the cleaning. The main advantage is that the central vacuum exhaust is normally outdoors and does not contribute to the redistribution of dust.

Plastic drop cloths are useful in many areas and, being good electrostatic collectors, tend to attract the finer particles from the air.

The mopping of floors is undesirable, because normal application by the average worker only redistributes the debris over a wider surface. Wet vacuuming, after a dry vacuum, is a preferred method for floors.

ACCESSORY CONTROLS

Bioclean rooms or controlled environment facilities by virtue of their design are positive-pressure devices. The opposite situation occurs in an infectious disease research laboratory or a facility for housing infected animals. In bioclean facilities, negative pressure is employed to prevent escape of potentially infected material. The following discussion is related only to positive-pressure facilities, but will identify the problem areas for both situations.

Sufficient pressure should be utilized to compensate for leaks, door openings, and the like. The purpose, of course, is to have contaminants flowing in an outward or opposite direction from that normally experienced in the design of infectious disease research laboratories. The same pressure required for this (usually 0.2 to 0.4 in. water gauge) works in opposition in opening doors, keeping doors closed, fire control, and filtering. To maintain air flow at 40 to 150 cfm, a large system will be required.

It is necessary to monitor the pressure drop across the prefilters to signal excessive loading. Manometers are generally used, because air velocity can also be derived and an instant evaluation of the whole system made. Carefully kept logs of pressure drops indicate variations from normal and the need to clean or change filters. Intermediate air locks generally precede the clean room to alleviate the pressure differential by stepping down pressure. These areas generally have an independent air handling system and monitoring manometers.

Heat or smoke detectors are mandatory in facilities where high volumes of air are handled. Catastrophe comes quickly in these events, so the system must be both sensitive and reliable. Temperature and humidity are routinely monitored manually for access areas and continuously by graphic recorder for the bio-clean room.

ACCEPTANCE

Assuming all preacceptance procedures have been followed, the actual acceptance is straightforward. It is based primarily on particulate requirements rather than microbiological considerations. Microbiological acceptance protocols may be desirable for liquid waste treatment systems, etc., but, in general, it is difficult to quantitate or specify materials or construction techniques on a strictly microbiological basis. If proper use has been made of knowledgeable engineering and biological personnel on the planning team, appropriate physical particulate loading parameters can be selected that correlate with the desired biological parameters.

The acceptance is usually based on three considerations, namely:

(1) Physical inspection
(2) Mechanical performance
(3) Meeting the appropriate standard for particulate matter.

PHYSICAL INSPECTION

A white glove inspection, while seemingly childish to some contractors, is a necessity. Plenums, ductwork, fan blades, floors, walls, and sills must be clean before start-up. Filters deserve particular attention. If a facility has been run after cleaning for several days without absolute filters in place, start-up should show both clean prefilters as well as absolute filters.

Other areas need close attention. An example of this is the tile work. Floor tile, as well as wall tile, should not show excess material in joints. Seams should be tight and, if necessary, coated to preclude harborage of debris and bacteria. Wall tiles should have the grouting glazed for easy cleaning. This is obviously important where pathogenic material is handled. Cove moldings that join floors and walls must be carefully installed to prevent harborage of physical and microbiological contaminants.

Painted surfaces should be examined for water blisters, cracks, flaking, and peeling. Any indications of these defects are positive indications of future contamination problems. Where surfaces have been reworked, direct supervision should be employed, because the tendency is to repaint rather than to correct the underlying defects.

In summary: The physical inspection of any controlled environment area should be directed at the small details and the subtle variations from normal. Any compromise from specifications should be considered a potential compromise of the intended use and therefore should be unacceptable.

MECHANICAL PERFORMANCE

Salient system features should be checked, performance curves examined, and performance monitored against the design drawings and specifications. Examples of some parameters that should be monitored are as follows:

(1) Air flow
(2) Pressure drops across filters
(3) Leaks (smoke or peppermint test) in sealed ductwork
(4) Setting of balancing valves, dampers, splitters, and other fixed setting controls
(5) Fan efficiency
(6) Temperature rise of all motors
(7) Pressure drop distribution through the facility
(8) Humidity
(9) Makeup air control sensitivity
(10) Illumination levels
(11) Fire control and alarm sensitivity
(12) Air-conditioning balance and capacity
(13) Specified temperature range in each zone
(14) Correct filter installation

COMPLIANCE AND STANDARDS

The ultimate requirement in a positive-pressure contamination controlled environment is to demonstrate the limit of maximum particulate concentration. Three primary methods of determining this concentration are discussed in the following paragraphs.

Ultrahigh Efficiency Filter Leak Check

This test procedure, which has been described in detail,[3] is performed by introducing smoke particles of a size equal to or larger than those specified for the filter system. These particles are introduced upstream from the filter bank, and the filters, frame gaskets, and mating surfaces monitored downstream for leaks. A light scattering-photometer type of instrument is the required means of detection.

Particulate Monitoring

This is a longer-term test, whereby samples of air from the facility are monitored for several days for particulate counts. Monitoring should be conducted at the critical operation location and upstream from the operator or work area. Care must be exercised, since contamination from the operator could invalidate the test.

Microbiological Considerations

While it is feasible to introduce a nonpathogenic aerosol challenge before the ultrahigh-efficiency filter system, and counts made by any of the classical means, other means are more practical. It has been shown that the particulate standards that have been established for filter testing can be correlated with microbiological testing. An extensive evaluation of a commercial vaccine production laboratory,[7] describes tests that were designed to evaluate the microbiological hazards associated with equipment, general building design, construction operational features, effluent treatment system, and routine research operations in such a newly constructed facility. The methods and tests described should be evaluated in conjunction with the purpose and objective of the facility and appropriate microbiological tests employed to complement the physical acceptance tests that are selected.

CONCLUSION

It is felt that preconditioning, pre- or in-process inspection is the most important phase in achieving a safe, operational facility for microbial contamination control. The contract documents for such a project must clearly specify the acceptance procedures that will be employed upon completion of construction. The relationship between compliance with the acceptance procedures and final payment must be stipulated. Some portions of the acceptance procedures, particularly the microbiological evaluations, should be performed by either "in-house" personnel or by experienced consultants. Because of the complexity of these procedures, it may be necessary to require that the general contractor meet only the physical parameters, and to arrange for the performance of the biological evaluations by "in-house" personnel or by consultants to insure safe operation of the facility.

CHAPTER 6

Preventive Safety

INTRODUCTION AND GENERAL PREVENTIVE MEASURES

Any facility handling agents of known or questionable pathogenicity for man and laboratory animals should offer the best possible environmental control and containment for the following areas of concern: (1) protection of the investigator and all associated supporting personnel from all known and potential biohazards; (2) protection of the experiment, resource materials, and animals by elimination of all possible routes of cross-contamination and cross infection; and (3) protection of the exterior environment against discharge of hazardous material.[58]

Hazards encountered in the infectious disease laboratory may be caused by (1) the ignorant, careless, or indifferent worker; (2) poor laboratory practices; (3) poor architectural design or inadequate equipment; (4) handling of agents that are highly communicable under inadequate laboratory conditions; or (5) a combination of these causal factors.

The employment of safe techniques and biological monitoring procedures, installation of laboratory safety equipment, proper laboratory design, and the vaccination and instruction of laboratory personnel are ways of minimizing the risk of unwanted biological contamination. These factors have been reviewed.[20,80]

Safety Rules for Infectious Disease Laboratories

Personnel working in an infectious disease laboratory must follow sound and established rules to prevent unwanted microbiological contamination. The following list of rules is intended to serve only as a guideline. The responsible investigator is urged to establish a more specific list for each individual infectious agent under investigation in his laboratory. These safety rules for infectious disease laboratories are modified from Phillips.[99] A general understanding of these will be helpful to the designer of the facility in which they will be used.

(1) Only authorized employees, students, and visitors should be allowed to enter infectious disease laboratories or utility rooms and attics of biological containment laboratories.
(2) Food, candy, gum, or beverages for human consumption should not be taken into infectious disease laboratories.
(3) Smoking should not be permitted in any area in which work on infectious or toxic substances is in progress. Employees who have been working with infectious materials should thoroughly wash and disinfect their hands before smoking.
(4) Library books and journals should not be taken into rooms where work with infectious agents is in progress.
(5) An effort should be made to keep all other surplus materials and equipment out of these rooms.
(6) Drinking fountains should be the sole source of water for drinking by human occupants.
(7) According to the level of risk, the wearing of laboratory or protective clothing may be required for persons entering infectious disease laboratory rooms. Likewise, showers with a germicidal soap may be required before exit.
(8) Contaminated laboratory clothing should not be worn in clean areas or outside the building.
(9) Before centrifuging, inspect tubes for cracks, inspect the inside of the trunnion cup for rough walls caused by erosion or adhering matter, and carefully remove bits of glass from the rubber cushion. A germicidal solution added between the tube and trunnion cup not only disinfects the outer surface of both of these, but also provides an excellent cushion against shocks that might otherwise break the tube.
(10) Avoid decanting centrifuge tubes. If you must do so, afterwards wipe off the outer rim with a disinfectant; otherwise the infectious fluid will spin off as an aerosol. Avoid filling the tube to the point that the rim ever becomes wet with culture.
(11) Water baths and Warburg baths used to inactivate, incubate, or test infectious substances should contain a disinfectant. For cold water baths, 70% propylene glycol is recommended.
(12) When the building vacuum line is used, suitable traps or filters should be interposed to insure that pathogens do not enter the fixed system.
(13) Deepfreezes and dry ice chests and refrigerators should be checked and cleaned out periodically to remove any broken ampules, tubes, etc., containing infectious material. Use rubber gloves and respiratory protection during this cleaning. All infectious or toxic material stored in refrigerators or deepfreezes should be properly labeled.

(14) Insure that all virulent fluid cultures or viable powdered infectious materials in glass vessels are transported, incubated, and stored in easily handled, nonbreakable, leakproof containers that are large enough to contain all the fluid or powder in case of leakage or breakage of the glass vessel.
(15) All inoculated Petri plates or other inoculated solid media should be transported and incubated in leakproof pans or other leakproof containers.
(16) Care must be exercised in the use of membrane filters to obtain sterile filtrates of infectious materials. Because of the fragility of the membrane and other factors, such filtrates cannot be handled as noninfectious until culture or other tests have proved their sterility.
(17) Develop the habit of keeping your hands away from your mouth, nose, eyes, and face. This habit may prevent self-inoculation.
(18) No person should work alone on an extremely hazardous operation.
(19) Broth cultures should be shaken in a manner that avoids wetting the plug or cap.
(20) Diagnostic serum specimens carrying a risk of serum hepatitis should be handled with rubber gloves.

Biological Monitoring of the Environment

The integrity of any microbiological barrier system must be checked periodically to evaluate its effectiveness. Common tests include microbial air sampling, particle size sampling, surface sampling, and surface contamination accumulation tests (Table 10). Biological monitoring of liquid wastes may be accomplished by procedures common to diagnostic bacteriology, mycology [4,54] and virology.[70] It has been found that methods employing ultracentrifugation for the recovery of viruses from sewage offer optimal recovery rates.[49,76] The common methods employed for the sampling of microbial aerosols have been summarized.[136] These include high- and low-velocity impingement in liquids, impaction on solid surfaces, filtration, centrifugation, electrostatic precipitation, and thermal precipitaton.

The types of air samplers available have been reviewed.[1,9] Specific sampling procedures have been presented for air filters;[27,28,109,119,122] for electrostatic precipitators;[27] and for an electric air sterilizer.[29]

Personnel Monitoring

Monitoring of laboratory personnel is an essential part of any microbiological monitoring system. Such a system should include: (1) regular serological and microbiological culture tests, biochemical function studies, and chest X-rays; and (2) the necessary vaccination and prophylactic measures.

TABLE 10. Microbiological Sampling of Environment

Object of Test	*Methods Employed*	*Type of Data Obtained*	*References to Equipment*
Microbial air sampling	Air impaction samplers	Viable organisms/cu ft of air	Henderson (1952) Druett & May (1952)
	Liquid impingers	Microorganisms/cu ft of air	Decker & Wilson (1954)
	Settling plates	Viable organisms/sq ft/hr	Tyler & Shipe (1959) Shipe et al. (1959) Tyler et al. (1959)
Particle size sampling	Liquid impinger with pre-impinger Anderson sampler Single-stage impactor device Casella cascade impactor	Relative proportions of different sized particles/cu ft of air	Anderson (1958) Lidwell (1959) Mitchell & Pilcher (1959) May & Druett (1953) Malligo & Idoine (1964) Sonkin (1950)
Surface sampling	Vacuum probe Cotton swabs Rodac plates	Microorganisms/unit of surface area	Hall & Hartnett (1964) Sandia Corp. (1967)
Surface contamination accumulation tests	Small sterile strips of stainless steel, glass or plastic	Viable organisms/unit of time	Portner et al. (1965)
Analyses of liquid waste and other fluids	Diagnostic methods of bacteriology, mycology and virology	Number and type of organisms/unit volume	Bailey & Scott (1962) Harris & Coleman (1963) Lennette & Schmidt (1964)

Attempts have been made to include in health monitoring programs any laboratory tests that might predict the possible onset of disease long before the appearance of any clinical symptoms. With some diseases, such as leukemia and other malignancies, however, detailed information is lacking on preclinical stages. For this reason, medical monitoring in these areas usually consists of no more than a preemployment physical examination, storage of a preemployment serum sample, and perhaps storage of serum samples taken at irregular intervals or following a laboratory accident or frank illness.[58] Failure to obtain preemployment serum samples in anticipation of their future usefulness and irreplacability is generally considered to be a serious omission. Despite a general lack of identification of those factors that might prove to be useful for detecting preclinical changes for certain diseases, the clinical medical field offers some logical choices which deserve critical evaluation as tools in a personnel medical monitoring program.

Laboratory Safety Equipment

Safety Cabinets. To date, the most widely accepted piece of equipment used for the control of microbiological contamination is the safety cabinet. These units have been in use for several years at various installations where highly infectious agents are handled. The two basic designs for these cabinets, the partial and absolute barrier, have been discussed previously.

Gastight cabinets can be modified for many specialized functions. They are used as containment devices or enclosures for back-mounted incubators and bottom-mounted refrigerators, freezers, and centrifuges when these items are used with agents that are potentially hazardous and cannot be handled by conventional equipment in the open laboratory.

Another advantage of the gastight cabinets is that they can be combined to form gastight modular cabinet systems,[50] thus enabling investigators to carry out a series of operations under total containment. Entrance and exit from such a system is maintained under a negative pressure of approximately one inch of water.

Partial barrier cabinets include the conventional inward flow units with turbulent air flow at the work surface, and the newer laminar downflow units. These latter units provide streamlined air movement throughout and simultaneously offer both personnel *and* product protection. Such laminar flow devices will undoubtedly find application for all but the most hazardous microbiological operations.

Animal Holding Equipment. Infected animals can be housed in cages with solid bottoms and sides and screen tops. In this situation the cage racks should be equipped with ultraviolet light to prevent the spread of infectious agents. In some instances, it may be desirable to house infected animals in a modular cabinet system. Infected animals may also be housed in individually sealed cages equipped with filters and ventilated through an air exhaust system.[20,62,63,94] Aerosol chambers and transfer hoods employed in aerosol exposure of laboratory animals have also been described.[62]

Safety Devices

Pipetting Devices. A considerable number of laboratory-acquired infections have been caused by pipetting accidents, particularly those involving mouth pipetting. Such accidents can be eliminated to a great extent by the use of safety pipetting devices. Table 11 lists several commercially available pipetting devices. The following pipetting rules should be established in the infectious disease laboratory:

(1) No infectious or toxic materials should be pipetted by mouth.
(2) No infectious mixtures should be prepared by bubbling expiratory air through a liquid with a pipette.
(3) No infectious material should be blown out of pipettes.
(4) Pipettes used for the pipetting of infectious or toxic materials should be plugged with cotton.
(5) Contaminated pipettes should be placed horizontally in a pan containing enough suitable disinfectant to allow complete immersion. They should not be placed vertically in a cylinder. The pan and pipettes should be autoclaved as a unit and replaced by a clean pan with fresh disinfectant.
(6) Infectious material should not be mixed by mouth pipetting.

Syringes.

(1) Only syringes of the needle-locking (Luer-Lok) type should be used with infectious materials.
(2) Use an alcohol-soaked pledget around the stopper and needle when removing a syringe and needle from a rubber-stoppered vaccine bottle.
(3) Expel excess fluid and bubbles from a syringe vertically into a cotton pledget soaked with disinfectant, or into a small bottle of cotton.
(4) Before and after injection of an animal, swab the site of injection with a disinfectant.

Safety in Centrifuging. If centrifuging of infectious material is carried out in an open laboratory area, safety centrifuge cups (Figure 7) should be used to prevent spread of infectious aerosols in the event of tube breakage. Removal of material from the individual centrifuge cups should be done in a safety cabinet.

TABLE 11. Safety Pipetting Devices

Trade Name	*Manufacturer*
Accropet	Manostat Corporation New York, New York
Adams Micropipettor	Clay-Adams Inc. New York, New York
Biopette	Bioquest Cockeysville, Md.
Caulfield Pipettor	Caulfield Safety Devices Philadelphia, Pennsylvania
Clinac Pipettor	LaPine Scientific Company Chicago, Illinois
Digi-Pet	J. H. Berge, Inc. S. Plainfield, New Jersey
Filler, Pipet	The Nalge Company Rochester, New York
Fisher Pipettor	Fisher-Scientific Company Pittsburgh, Pennsylvania
Kadavy Micropipettor	A. S. Aloe Company St. Louis, Missouri
Micropipettor	Alfred Bicknell Associates Cambridge, Massachusetts
Plasti Pet	Bel-Art Products Pequannock, New Jersey
Propipette	Schaar and Company Chicago, Illinois
Pumpett	LaPine Scientific Company Chicago, Illinois
Vadosa	Bel-Art Products Pequannock, New Jersey

If a small tabletop model centrifuge is being used to centrifuge infectious materials, then the centrifuge should be operated in a ventilated cabinet. It is best to have the glove panel in place and the glove ports covered, since the operating centrifuge may create strong air currents that could cause escape of infectious materials.

Safety Blendors. The kitchen type of high-speed blendor is often used in the microbiological laboratory for mixing various materials including materials containing infectious agents. This blendor has been shown to produce aerosols and has been implicated in the laboratory infection of several workers. A leakproof high-speed mixing bowl in which blended materials are removed under a vacuum and aseptic conditions has been described.[104] With this device, infectious materials may be blended in the open laboratory without dissemination of infectious aerosols. This blendor is manufactured by the Waring Blendor Company as a "Biological Dispersal Unit" or "Aseptic Dispersal" and is distributed by most of the scientific equipment supply companies.

DISINFECTION AND STERILIZATION

All infectious or toxic materials, equipment, or apparatus should be autoclaved or otherwise sterilized before being washed or disposed of. Each individual working with infectious material should be responsible for its sterilization before disposal. Infectious or toxic materials should not be placed in autoclaves overnight in anticipation of autoclaving the next day.

To minimize hazard to firemen or disaster crews, at the close of each work day, all infectious or toxic material should be (1) placed in the refrigerator, (2) placed in the incubator, or (3) autoclaved or otherwise sterilized before the building is closed. Autoclaves should be checked for operating efficiency by the frequent use of Diack, or equivalent, controls.

All laboratory rooms containing infectious or toxic substances should designate separate areas or containers labeled: INFECTIOUS—TO BE AUTOCLAVED or NOT INFECTIOUS—TO BE CLEANED. All infectious disease work areas, including cabinetry, should be prominently marked with a biohazards warning symbol.[5]

Floors, laboratory benches, and other surfaces in buildings in which infectious substances are handled should be disinfected with a suitable germicide as often as deemed necessary by the supervisors. After completion of operations involving plating, pipetting, centrifuging, and similar procedures with infectious agents, the surroundings should be disinfected.

Floors drains throughout the building should be flooded with water or disinfectant at least once each week in order to fill traps and prevent backing up of sewer gases. Floors should be swept with push brooms only. A floor-sweeping compound is effective in lowering the number of airborne organisms. Water used to mop floors should contain a disinfectant. Elimination of sweeping through use of dry and/or wet pick-up vacuum cleaners with high-efficiency exhaust air filters is recommended.

Stock solutions of suitable disinfectants should be maintained in each laboratory. All laboratories should be sprayed with insecticides as often as necessary to control flies and other insects. Vermin-proofing of all exterior building openings should be given close attention.

No infectious substances should be allowed to enter the building drainage system without prior sterilization.

Mechanical garbage disposal units should not be installed for use in disposing of contaminated wastes. These units release considerable amounts of aerosol.

LABORATORY ANIMALS

Animal Cage Labeling

All animal cages should be marked to indicate the following information:

(1) Uninoculated animals.
(2) Animals inoculated with noninfectious material.
(3) Animals inoculated with infectious substances.
(4) Other pertinent information.

Cages Housing Infected Animals

Cages used for infected animals should be cared for in the following manner:

(1) Careful handling procedures should be employed to minimize the dissemination of dust from cage refuse and animals.
(2) Cages should be sterilized by autoclaving. Refuse, bowls, and watering devices will remain in the cage during sterilization.
(3) All watering devices should be of the non-drip type.
(4) Each cage should be examined each morning and at each feeding time so that dead animals can be removed.

Handling Infected Animals

(1) *Special attention* should be given to the humane treatment of all laboratory animals in accordance with the Principles of Laboratory Animal Care as promulgated by the National Society for Medical Research and existing laws or regulations.[68]
(2) Monkeys should be tuberculin-tested and examined for herpetic lesions.
(3) Persons regularly handling monkeys should receive periodic chest X-ray examination and other appropriate tuberculosis detection procedures.

(4) When animals are to be injected with pathogenic material, the animal caretaker should wear protective gloves and the laboratory workers should wear surgeon's gloves. Every effort should be made to restrain the animal to avoid accidents that may result in disseminating infectious material.
(5) Heavy gloves should be worn when feeding, watering, or removing infected animals. Under no circumstances should the bare hands be placed in the cage to move any object.
(6) Animals in cages with shavings should be transferred to clean cages once each week unless otherwise directed by the supervisor. If cages have false screen platforms, the catch pan should be replaced before it becomes full.
(7) Infected animals to be transferred between buildings should be placed in aerosol-proof containers.

Animal Rooms

(1) Doors to animal rooms should be kept closed at all times except for necessary entrance and exit.
(2) Unauthorized persons should not be permitted entry to animal rooms.
(3) A container of disinfectant should be kept in each animal room for disinfecting gloves, boots, and general decontamination. Floors, walls, and cage racks should be washed with disinfectant frequently.
(4) Floor drains in animal rooms should be flooded with water or disinfectant periodically to prevent backing up of sewer gases.
(5) Shavings or other refuse on floors should not be washed down the floor drain, if these are present.
(6) An effective poison should be maintained in animal rooms to kill escaped rodents.
(7) Special care should be taken to prevent live animals, especially mice, from finding their way into disposable trash.

Necropsy of Infected Animals

(1) Necropsy of infected animals should be carried out in ventilated safety cabinets.
(2) Rubber gloves should be worn when performing necropsies.
(3) Surgeon's gowns should be worn over laboratory clothing.
(4) The fur of the animal should be wet with a suitable disinfectant.
(5) Animals should be pinned down or fastened on wood or metal in a metal tray.
(6) Upon completion of necropsy, all potentially contaminated material should be placed in suitable disinfectant or left in the necropsy tray. The entire tray should be autoclaved at the conclusion of the operation.

(7) The inside of the ventilated cabinet and other potentially contaminated surfaces should be disinfected with a suitable germicide.
(8) Grossly contaminated rubber gloves should be cleaned in disinfectant before removal from the hands, preparatory to sterilization.
(9) Dead animals should be placed in proper leakproof containers and thoroughly autoclaved before being placed outside for removal and incineration.

Conclusions

Techniques of microbiological contamination control have been employed for many years. Lord Joseph Lister became one of the earliest practitioners when, in the 1870's, he sprayed carbolic acid in operating rooms to prevent surgical sepsis. Today microbial contamination control finds application in many diverse areas. Moreover, the direction of man's science and technology signals an ever-increasing need for the control of microbial contaminants through proper facility design. The ultimate in microbiological contamination control is found in the achieving and maintenance of sterility, but this condition is usually definable only in statistical terms on the basis of a sample of the treated population and on the known characteristics of microbes in their reaction to inactivating treatments. Man's reaction to a microbiologically contained system is such that control is better maintained if he manipulates the system remotely; thus, enclosure of the technique or operation is more convenient than the enclosure of man himself.

The general principles of design of facilities for microbial contamination control and the requirements for measurement and testing discussed herein illustrate the need for the involvement of trained microbiologists, architects, and engineers in an interdisciplinary approach in every control endeavor. They also illustrate the need for additional summarized information on the standards of control, the techniques required to achieve control, and means of microbiologically determining when control is reached. Most, if not all, microbiological contamination control efforts should contain elements from the five stages shown in Table 2. Finally, it must be realized that the entire field of microbial contamination control is dynamic and that new techniques and new solutions will be required as new problem areas arise.

Glossary

Absolute Containment Capacity for keeping or retaining completely any specified substance.

Aerosol A colloid of liquid or solid particles suspended in a gas, usually air.

Agent Any power, principle, or substance capable of acting upon the organism, whether curative, morbific, or other.

Air Incinerator An electric or fuel-fired furnace for the sterilization of microbiologically contaminated air (or other gases).

Air Lock An unventilated space isolated by doors, used to separate areas with different levels of contamination and at different air pressures, which permits passage of personnel and/or equipment without air flow. (See also **UV Air Lock.**)

Andersen Sieve Sampler A sampler consisting of six impaction stages positioned in series, each containing a plate perforated with a pattern of 400 holes. A petri dish containing agar is placed beneath each stage. The diameter of the holes is constant for each stage, but is smaller in each succeeding stage. Therefore, as air is drawn through the sampler, the jet velocity is uniform for each stage, but increases with each succeeding stage. The result is particles are collected on the agar plates in a distribution according to size. The distribution is as follows:

Stage	Hole Diameter (mm)	Jet Velocity (ft/sec)	Size Distribution [a] (microns)
1	1.81	3.54	8.2+
2	0.91	5.89	5-10.5
3	0.71	9.74	3-6
4	0.53	17.31	2-3.5
5	0.34	41.92	1-2
6	0.25	76.40	[b]—1

[a] Distribution based on smooth, spherical particles of unit density, 95% or more of particles collected.

[b] Lower range of sixth not known.

Animal Cage Rack Stack of steel shelves, generally movable, used to hold animal cages; sometimes equipped for UV irradiation and sometimes provided with exhaust manifold to accommodate ventilated cages.

Animal Holding Room Room meeting suitable isolation criteria used to house animals in cages before and after experimental use; may be in noncontaminated or contaminated areas.

Anisokinetic Sampling The taking of air samples under conditions in which the velocity and direction of the air entering the sampler are different from the velocity and direction of the ambient air. When sampling velocity is greater than that of the ambient air, smaller particles are selectively collected over larger particles, and vice versa, resulting in counting errors proportional to the particle size and distribution. (See also **Isokinetic Sampling**.)

Antiseptic A compound that prevents the multiplication of microorganisms. Bacteriostatic in action, not bactericidal.

Aseptic Technique The performance of a procedure or operation in a manner that prevents the introduction of septic material.

Attic An important utility service area for the laboratories containing much service equipment.

Autoclave (Steam Sterilizer) A chamber used for heat sterilization of materials and equipment by direct exposure to steam under pressure. Standard conditions for laboratory glassware are 15 lbs/in.^{2} (250° F; 121° C) for 15 to 20 minutes.

Backflow Preventer A manufactured piping device of the type that has two spring-loaded vertical check valves and one spring-loaded, diaphragm-activated differential pressure relief valve. It is installed in a water supply line to prevent reversal of water flow in case the supply pressure falls below the downstream pressure. (See also **Break Tank** and **Vacuum Breaker**.)

Bactericide An agent which kills vegetative bacteria and some spores; synonymous with germicide.

Bacteriostat An agent which stops the growth and multiplication of bacteria, but does not necessarily kill them. Usually growth resumes when the bacteriostat is removed.

Biological Safety Cabinet A ventilated enclosure that provides a physical barrier between a worker and a hazardous operation. It may be used with an open front (or open glove ports) and a high rate of ventilation away from the operator, like a fume hood, or with a closed front and attached rubber gloves. In the latter use, protection depends upon a negative pressure maintained within the cabinet. The ventilation air exhausts through a high-efficiency filter. A gastight, ventilated safety cabinet maintained at negative pressure, in which work is done using attached rubber gloves should be used for highly infectious material. Ultrahigh-efficiency filters on the inlet and exhaust, and provision for exhaust air incineration is required.

Biological Spill Alarm A system provided in large infectious disease buildings to warn building occupants that release of hazardous material has occurred. Alarm switches are conveniently located throughout the building and give a coded audible signal and actuate a warning light.

Break Tank A tank that provides an air space in a water supply line in such a manner as to prevent reversal of water flow in case the supply pressure falls below the downstream pressure. It is considered more positive than the **Backflow Preventer** or **Vacuum Breaker**.

Cabinet Array A number of absolute barrier biological safety cabinets joined together. An array may be divided into two or more **Cabinet Systems** by gastight doors or fixed partitions.

Cabinet System A number of absolute barrier biological safety cabinets joined together to provide a single space, with a single inlet and exhaust for ventilation.

Casella Slit Sampler A solid medium impactor consisting of a chamber in which an agar plate revolves beneath one or more slits. Two models of the Casella sampler are commercially available, one that collects at 28 liters per minute, and one that collects at 700 liters per minute. Both models operate at a choice of three rotation speeds for the agar plate, one revolution per ½ minute, per two minutes, and per five minutes.

Change Room(s) Grouping of dressing rooms, locker rooms, lavatories, air locks, and showers to provide personnel access to and egress from contaminated areas without allowing escape of contamination.

Clean Change Room Dressing room for removal of street clothes and donning of laboratory-type clothing before entering contaminated change room, through an air lock.

Colony A discrete accumulation of microbial growth on the surface of a solid culture medium which is visible to the naked eye. A colony can be the result of the growth and multiplication of a single microbial cell, or of a clump of two or more cells. A colony formed from a single cell is referred to as a clone, but it is usually not possible to differentiate by mere visual examination a clone from a colony that may have formed from, for example, 100 cells.

Contaminated Area A building area with definite boundaries where hazardous biological work is being carried out, separated from noncontaminated and other contaminated areas by suitable barriers.

Contaminated Change Room Dressing room for removal of laboratory-type clothing before entering clean change room, through a mandatory shower, to don street clothing.

Contaminated Service A service or utility, such as water or vacuum, which serves a contaminated area and is therefore segregated from similar services to noncontaminated areas, even though the service itself is noncontaminated.

Critical Orifice An opening through which air passes, the diameter of which is such that the velocity of the air is independent of the pressure differential across the opening, provided the differential is above a certain value. A number of aerosol samplers operate with a critical orifice, notably the AGI and TDL samplers. When the samplers are operated at or above a certain minimum vacuum, sampling rate is unaffected by fluctuations in the vacuum.

Culture Media A solid or liquid substance containing nutrient materials for the cultivation of microorganisms.

D-Value The decimal reduction time. The sterilization time necessary to reduce a microbial population by 90% (one log). It must be assumed that the rate of kill is linear with respect to the logarithm of the surviving population, then:

$$D = \frac{t}{(\log A - \log B)}$$

where t=time of sterilization; A=population at time A; B=population at time $A+t$. The reciprocal of the D value $(1/D)$ is referred to as the death rate constant, k.

Decontamination The destruction or removal of living organisms to some lower level, but not necessarily to zero. (See also **Sterilization**.)

Deep-bed Filter Common form of high-efficiency filter for low-pressure use in ventilation systems.

Diaphragm Valve Widely used in contaminated service because of zero leakage at the stem.

Disinfectant A chemical agent that kills vegetative bacteria, fungi, and viruses, but not spores.

Disinfectant Shower Unit at exit from ventilated suit area in which the suit is externally decontaminated by mist or spray of disinfectant, such as peracetic acid, before being removed.

Droplet An airborne particle consisting primarily of liquid. Droplets are usually generated by talking, sneezing, or coughing, and while some settle out almost immediately, many dry to become droplet nucleii to add significant numbers of microorganisms to the air.

Droplet Nuclei Airborne particles that originated as droplets, but have dried and left behind mucus, salt, microorganisms, and other material that may have been associated with the droplet. Because of their small size, they frequently remain suspended in the air for long periods of time.

Dry Heat Sterilization Thermal sterilization carried out in the absence of added moisture. For the same sterilization time, dry heat usually requires a higher temperature than moist heat to achieve the same degree of sterilization.

Dunk Bath A tank device containing a liquid germicide that allows materials to be passed into and out of a gastight enclosure without transfer of air or microorganisms.

Enclosure A general term that includes cabinets and other devices designed to isolate or separate a working area for product protection or personnel protection reasons.

Ethylene Oxide A colorless gas having remarkable penetrating power that is frequently used as a sterilizing or decontaminating agent. It is effective at room temperature, although a certain level of moisture is required for optimum effectiveness. The flammability hazard is greatly reduced by diluting the gas with carbon dioxide or fluorocarbons such as Freon.

Exfiltration (Ventilation term.) Ductless flow of air from a space to an adjoining space at lower pressure.

Filter A device used for removal of undesirable particulates, including microorganisms, from air or other gases. Includes ultrahigh-efficiency filters and high-efficiency filters, as well as less efficient types.

Fomites Inanimate objects or materials that act as intermediate carriers of microbial contamination. Contaminated tools used to assemble a sterile spacecraft could be fomites. (Singular=fomes)

Gas Sterilizer An autoclave that is designed for or has been modified to permit optional use of a gaseous decontaminant instead of steam for sterilizing materials.

Germfree Free of all microbial life detectable by examination.

Gravity Exhaust (Ventilation term.) Discharge of air, resulting only from pressure differential, from a ventilated room to the outdoors through an exhaust duct.

Heat Shocking A procedure used in bacteriological assays to eliminate vegetative cells from a sample, but to allow spores to survive. Usually carried out by heating samples to 80°C for 20 minutes.

High-efficiency Filter Having a nominal efficiency of 99% for removal of 0.3 to 0.5 micron particles from air.

Hyperbaric Chamber A chamber containing more than the normal atmospheric pressure and an increased oxygen tension and used for the treatment of certain medical conditions.

Hypobaric Chamber (Altitude Chamber) A chamber containing less than the normal atmospheric pressure and used to simulate conditions at distances above the earth.

Infectious Microorganisms As used in this book, the term is restricted to microorganisms capable of producing a disease process in man or domestic animals.

Infiltration (Ventilation term.) Ductless flow of air into a space from an adjoining space at higher pressure.

Isokinetic Sampling The taking of air samples under conditions in which the velocity and direction of the air entering the sampler are the same as the velocity and direction of the ambient air. (See also **Anisokinetic Sampling.**)

Isolator A barrier, usually flexible plastic, that operates at a positive air pressure to seal off the internal work space from the surrounding environment.

Laminar Air Flow Air flow in which the entire body of air within a designated space moves with uniform velocity in a single direction along parallel flow lines.

Membrane Filter A filter medium made from various polymeric materials, such as cellulose, polyethylene, or tetrafluoroethylene. Membrane filters usually exhibit very narrow ranges of effective pore diameters, making the filters very useful in the collection and sizing of microscopic and submicroscopic particles, and in the sterilization of liquids.

Microbiological Barrier A material object or set of objects that separates, demarcates, or serves as a barricade to prevent the passage or migration of microorganisms.

Microbiological Tests Methods of examination of specimens, objects, or materials to determine the presence or absence of microorganisms, their taxonomic identification, and/or their relative frequency and types.

Micron A unit of length equivalent to 10^{-6} meters, or 10^{-3} millimeters, or approximately 1/25,000 inch. Symbol μ.

Microorganisms Single-celled plants or animals; organisms of microscopic or ultramicroscopic size. Commonly refers to fungi, bacteria, Rickettsia, and viruses.

Minimum Turbulence Air Flow Unidirectional air flow within a work space where undisturbed air moves in parallel lines and where the location of people and objects within the work space is controlled to allow reasonable recovery from turbulence downstream of the people and objects. The rate of air movement is usually such that surface contamination from airborne particulate fallout is minimized.

Noncontaminated Area An area in the laboratory building with definite boundaries designed to be free of harmful microorganisms.

Partial Containment An enclosure which is so constructed that contamination between its interior and the surroundings is minimized by the movement of air. A laminar flow clean bench is an example of a partial barrier in which contamination is minimized by the outward flow of air. A chemical hood is also a partial barrier in which contamination of the surroundings is minimized by the inward flow of air. (See also **Absolute Containment.**)

Planetary Quarantine An area of endeavor involved in the prevention of contamination of other planetary bodies and the moon by terrestrial organisms, as well as the prevention of contamination of Earth by organisms from extraterrestrial bodies.

Plenum When not otherwise specified, refers to filter chamber upstream of exhaust fan in the building ventilation system.

Receiving Room, Contaminated An area for holding contaminated equipment and materials until they can be sterilized and passed through double-door autoclaves or gas sterilizers that open into the noncontaminated receiving room.

Receiving Room, Noncontaminated A service room generally at the rear of the building that is maintained as a noncontaminated area. Supplies delivered to the building are placed in the receiving room before transfer through a UV air lock to the contaminated receiving room.

Refuse Incinerator A fuel-fired furnace for the combustion of organic wastes, in which all gases will have reached a minimum temperature of 1350°F before discharge.

Respirator A conventional device covering the nose and mouth, which provides a filter for inspired air.

Reyniers Slit Sampler The commercial version of the Fort Detrick "pot" or slit sampler. An impactor which collects airborne particles onto a 150 mm agar plate. The plate slowly rotates under the impaction slit, making it possible to obtain the microbial profile of a given air space over a period of one or two hours on a single agar plate. The sampler consists of a metal body and a spring wound clock mechanism which rotates the agar plate. The impaction slit width and its height above the agar surface are adjustable.

Rodac Plate A plastic culture plate so designed that when filled with sufficient agar culture medium, a meniscus is formed which rises above the rim of the plate. The plate can then be used to assay the microbiological contamination of smooth surfaces (tabletops, floors, skin) by pressing the raised agar against the surface. A number of microorganisms are transferred to the agar, and after suitable incubation visible colonies develop. The number of colonies appearing is a reflection of the degree of contamination of the surface.

Rodent-proof To incorporate prescribed structural and architectural features in building design that prevent access or harboring of rodents and other vermin.

Safety Shower Provided in chemical and radiological laboratories for same function as in conventional, nonbiological laboratories.

Servomechanism A mechanism within a barrier system controlled by a mechanism external to the system.

Speaking Diaphragm Plastic sheet installed in wall, door, or window to permit voice communication through barrier between areas of different levels of contamination.

Spore A body which some species of bacteria form within their cells which is considerably more resistant to unfavorable conditions (heat, germicides, ultraviolet light) than the vegetative cell. When favorable conditions return, the spore usually germinates, or transforms back to the vegetative form.

Sterility The state of being free from all living microorganisms.

Sterilization Complete destruction or removal of all living microorganisms.

TDL Sampler An impaction sampler similar in principle to the Casella slit sampler. Rotation rates for the agar plate can be selected from one revolution per ⅔ of a minute to one revolution per 180 minutes, depending on model selected.

Ultrahigh-efficiency Filter Having a minimum efficiency of 99.97% when tested with 0.3 micron "DOP" particles; generally employs pleated glass or asbestos paper. Also commonly referred to as "Absolute * filter," "HEPA filter," or "superinterception filter."

UV Air Lock An air lock located between areas of different levels of contamination and air pressure. It provides a dead air space for the transfer of personnel and/or equipment without air flow. The interior is irradiated with UV and painted with aluminum paint to give good UV reflectance.

UV Clothing Discard Rack A rack that holds a standard laundry bag and is protected at the top with a curtain of UV. Clothing worn in the contaminated laboratory is discarded into this laundry bag in the contaminated change room.

Ultraviolet Radiation (UV) Denoting the actinic or chemical rays beyond the violet end of the spectrum. In this book UV refers to the germicidal line at 2537 Å produced by mercury arc lamps.

Vacuum Breaker A device that is installed in a line or tank, where the breaker is not subjected to a downstream back-pressure, to prevent reversal of flow in case of accidental occurrence of an upstream suction.

Vegetative Cell A bacterial cell capable of multiplication, as opposed to the spore form, which cannot multiply. The vegetative form is usually considerably less resistant to adverse conditions, such as heat or germicides, than the spore form. (See also **Spore**.)

* Absolute ®—Cambridge Filter Corp., Lexington, Kentucky.

Ventilated Hood Hood covering entire head, pressurized with conditioned air by same hose system serving ventilated suits.

Ventilated Suit Pressurized outer garment, including head, hands, and feet, supplied by hose with conditioned air, for work in areas of high risk from infectious aerosols such as some animal rooms.

Viable Literally, "capable of life," and generally refers to the ability of microbial cells to grow and multiply as evidenced by, for example, the formation of colonies on an agar culture medium. Frequently organisms may be viable under one set of culture conditions and not under another set, making it extremely important to define precisely the conditions used for determining viability.

Viable Particle A discrete unit which is associated with one or more viable microorganisms. A viable particle may be an inert particle (dust, lint) on which is adhered one or more viable microorganisms, it may consist of an aggregate of two or more viable cells, or it may be simply a single viable cell. Most aerosol samplers, especially the liquid impingers, because of the severe stresses they impose on viable particles during collection, tend to break up viable particles. The results of most viable air sampling, therefore, indicate a contamination level somewhere between the number of viable particles in the air and the actual number of viable microorganisms in the air.

Viewing Panel Fixed window suitably sealed into an interior wall or door between two areas of different contamination level.

Viricide An agent that kills viruses.

Waste Collection Treatment Unit A unit for collecting and treating liquid waste, generally serving one building, consisting of a tank in which the contaminated liquid waste is collected and sterilized, either continuously or batch-wise.

References

1. Albrecht, J., "Quantitative Estimation of Bacterial Aerosols, *Zeit. fur Aerosol-Forsch und Therapie* **4,** 3–11 (1955).
2. Alschuler, J. H., "Air Treatment for Research Animal Housing," *Laboratory Animal Care,* **13,** No. 3 (Part 2, 1963).
3. Austin, P. R., and Timmerman, S. W., "Design and Operation of Clean Rooms," pp. 83–84, Business News Publishing Company, Detroit, Mich., 1965.
4. Bailey, W. R., and Scott, E. C., "Diagnostic Microbiology, C. V. Mosley, St. Louis, 1962 (327 pp.).
5. Baldwin, C. B., and Runkle, R. S., "Biohazard Symbol: Development of a Biological Hazards Warning Signal," *Science,* **158** (3798), 264–265 (1967).
6. Balzam, N., "Aseptische zucht vor tieren: I. Apparatus and Methods. II. Aseptische zucht von hakenworm mit vitaminhaltigem und vitaminfreier futter," *Acta. Biol. Exptl.* (*Warsaw*), **11,** 43–56; *Ber. Biol.,* **48,** 545 (1937).
7. Barbeito, M. S., "Biological Evaluation of a Commercial Vaccine Production Laboratory," Technical Report 65, Fort Detrick, Frederick, Md., May, 1965.
8. Barrett, J. P., Jr., "Sterilizing Agents for Lobund Flexible Film Apparatus," *Proc. Animal Care Panel,* **9,** 127–133 (1959).
9. Batchelor, H. W., "Aerosol Samplers," *Adv. Appl. Microbiol,* **2,** 31–64 (1960).
10. Berthelot, M., "Fixation directe de l'azota atmospherique libre par certains terrains argileux," *Compl. Rend. Acad. Sci.,* **101,** 775–784 (1885).
11. Bertin, R. J., "A Study of Contamination Control in Pharmaceutical Compounding with a Laminar Flow Clean Work Bench," A report submitted to the Graduate School, University of Minn., in partial fulfillment for req. of Master of Science in Hospital Pharmacy, 1966.
12. Billingsley, J. C., and LeVora, N., "A Self Contained Air System and Protective Suits," *Aerospace Medicine,* **34,** No. 12 (Dec. 1963).
13. Blenderman, L., "Water Service for Research Laboratories," *Air Conditioning, Heating and Ventilating,* (Oct. 1956).
14. Blickman, B. I., and Lanahan, T. B., "Ventilated Work Cabinets Reduce Lab Risks," *Safety Maintenance,* **120,** 34–36, 44–45 (1960).

15. Brewer, N. R., "Estimating Heat Produced by Laboratory Animals," *Heating, Piping and Air Conditioning,* pp. 139–141 (Oct. 1964).
16. Bruch, C. W., "Decontamination of Enclosed Spaces with Beta Propiolactone Vapor," *Am. J. Hyg.,* **73,** 109 (Jan. 1961).
17. Bruch, C. W., and Koesterer, M. G., "The Microbicidal Activity of Gaseous Propylene Oxide and its Application to Powdered or Flaked Foods" (1961).
18. Bruch, C. W.; Koesterer, M. G., and Bruch, M. K., "Dry Heat Sterilization: Its Development and Application to Components of Exobiological Space Probes," *Developments in Industrial Microbiology,* **4,** Washington, D. C., American Institute of Biological Sciences, 1963.
19. Bulloch, W., "The History of Bacteriology," Oxford Univ. Press, New York, 1938 (422 pp.).
20. Chatigny, M. A., "Protection Against Infection in the Microbiological Laboratory: Devices and Procedures," *Adv. Appl. Microbiol,* **3,** 131–192 (1961).
21. Chatigny, M. A. (Personal Communication), U.S. Naval Biological Laboratories, Oakland, Calif.
22. Cohendy, M., "Experiences sur la vie sons microbes," *Pasteur,* **26,** 106–137 (1912).
23. Cook, R. O., "Development of a Mechanically Ventilated Isolation Cage," *Applied Microbiology,* **16,** No. 5, 762–771 (1968).
24. Cook, R. O., "New Ventilated Isolation Cage," *Appl. Microbiol,* **16,** 762–771 (1968).
25. Corriell, L. L.; McGarrity, G., and Blakemore, W. S., "Studies on HEPA Filtered Vertical Flow Air in the Microbiological Laboratory and Operating Room," Proceedings, Sixth Annual Technical Meeting, American Association for Contamination Control, May 15–18, 1967.
26. Crossman, R. F., and Elsea, R., "Air Conditioning Laboratory Animal Quarters," *Air Conditioning, Heating and Ventilating* (July 1961).
27. Decker, H. M.; Geile, F. A.; Moorman, H. E., and Glick, C. A., "Removal of Bacteria and Bacteriophage from the Air by Electrostatic Precipitators and Spun Glass Filter Pads," *Heating, Piping and Air Conditioning* **23,** 128–38 (1961).
28. Decker, H. M.; Geile, F. A.; Harstad, J. B., and Gross, N. H., "Spun Glass Air Filters for Bacteriological Cabinets, Animal Cages and Shaking Machine Containers," *J. Bacteriol.* **63,** 377–83 (1952).
29. Decker, H. M.; Citek, F. J.; Harstad, J. B.; Gross, N. H., and Piper, F. J., "Time Temperature Studies of Spore Penetration Through an Electric Air Sterilizer," *Appl. Microbiol,* **2,** 33–36 (1954).
30. Decker, H. M.; Buchanan, L. M.; Hall, L. B., and Goddard, K. R., "Air Filtration of Microbial Particles," Public Health Service Publ. No. 953, U.S. Govt. Printing Office, Washington, D.C., 1962 (43 pp.).
31. Devereux, R. C. deB, and Charlton, R., "Design of a Pharmaceutical Laboratory," *Inst. Heating Ventilation Eng. J.,* **30,** 45–49 (1962). (Welcome Foundation, Ltd., England.

32. Dickens, F., and Jones, H. E. H., "Carcinogenic Activity of a Series of Reactive Lactones and Related Substances," *Brit. J. Cancer,* **XV,** No. 1, pp. 85–100 (March 1961).
33. Dolowy, W. C., "Medical Research Laboratory of the University of Illinois," *Proc. Animal Care Panel,* **11,** 267–280 (1961).
34. Doxie, F. E., and Ullom, K. J., "Human Factors in Designing Controlled Ambient Systems," *The Western Electric Engineer,* **XI,** No. 1 (Jan. 1967). (© 1967 by Western Elec. Co., Inc.)
35. Dwyer, J. L., "Contamination Analysis and Control," Reinhold Book Corp., New York, 1966.
36. Dugan, V. L.; Whitfield, W. J.; McDade, J. J. Beakley, J. W.; and Oswalt, F. W., "A New Approach to the Microbiological Sampling of Surfaces: The Vacuum Probe Sampler," Sandia Corp. Publication SC–RR–67–114, 1967.
37. Ernst, R. P., and Kretz, A. P., Jr., "Compatability of Sterilization and Contamination Control with Application to Spacecraft Assembly," *Journal of American Association for Contamination Control,* 1967.
38. Faust, F. R., "Cagewashing Area—Planning and Design," Proceedings of the Symposium on Research Animal Housing, *Laboratory Animal Care,* **13,** 221–467 (1962).
39. Favero, M. S. (Chairman); McDade, J. J.; Robertsen, J. A.; Hoffman, R. K.; and Edwards, R. W., "Microbiological Sampling of Surfaces," Technical Report, Biological Contamination Control Committee, American Association for Contamination Control, March 1967.
40. Fox, D. G., "An Empirical Study of the Application of a Horizontal Unidirectional Airflow System for a Hospital Operating Room," A Thesis submitted to the Faculty of the Graduate School of the Univ. of Minn., July 1967.
41. Fox, G., "Design of Clean Rooms," A Classified List of Selected References, 1955–1963, Division Research Services, National Institutes of Health, USPHS, Bethesda, Md., 1963 (15 pp., 127 ref.).
42. Fox, G. "Specific Pathogen Free Animal Production—Selected References 1931–1964," Research Facilities Planning Branch, Division Research Services, National Institutes of Health, April 1964.
43. Fricke, W., "Schutamassahme bei Bakteriologischem und Serologischem Arbeiten," Gustav Fischer, Jena, Germany, 1919.
44. Gilbert, G. L.; Gambill, V. M.; Spiner, D. R.; Hoffman, R. K., and Phillips, C. R., "Effect of Moisture on Ethylene Oxide Sterilization," *Appl. Microbiol.,* **12,** 496–503 (1964).
45. Glimstedt, G., "Bakterienfreie meerschweinchen," *Acta Pathol. Microbiol. Scand., Suppl.,* **30,** 1–295 (1936).
46. Goddard, K. R., "100% Outside Air Less Costly Than Recirculated Air," *Air Engineering* **1962**.
47. Gohr, F. A., "Hospital Engineering. Environmental Health and Safety," *Air Cond. Heat. and Vent.,* **61,** 52–58 (July 1964).

48. Graham, W. R., and Feenstra, E. S., "A Program for the Development of Pathogen Free Laboratory Animals," *Proc. Animal Care Panel,* Nov. 7–9 (1957).
49. Gravelle, C. R., and Chin, T. D. Y., "Enterovirus Isolations from Sewage: A Comparison of Three Methods," *J. Infect. Dis.,* **109,** 205–09 (1961).
50. Gremillion, G. G., "The Use of Bacteria-Tight Cabinets in the Infectious Disease Laboratory," *In* Proc. 2nd Symp. Gnotobiotic Technology, Univ. of Notre Dame Press, Notre Dame, Ind. pp. 171–182, 1959.
51. Gus, L., "Plumbing Design for an Aerospace Medical Research Center," *Air Cond. Heat. and Vent.,* Sept. 1961.
52. Gustafsson, B. E., "Lightweight Stainless Steel Systems for Rearing Germ-free Animals," *Ann. N.Y. Acad. Sci.,* **78,** 17–28 (1959).
53. Hall, L. B., and Hortnell, M. J., "Measurement of Bacterial Contamination on Surfaces in Hospitals," *Public Health Reps.,* **77,** 1021–1024 (1964).
54. Harris, A. H., and Coleman, M. B. (eds.), "Diagnostic Procedures and Reagents," (4th ed.), American Public Health Association, New York, 1963 (888 pp.).
55. Harris, G. J.; Gremillion, G. G., and Towson, P. H., "Test New Electric Incinerator Design for Sterilizing Laboratory Air," *Heating, Piping and Air Conditioning,* **36,** 94–95 (1964).
56. Harstad, J. B.; Decker, H. M.; Buchanan, L. M.; and Miller, M. E., "Air Filtration of Submicron Virus Aerosols," Proceedings, Sixth Annual Technical Meeting, American Association for Contamination Control, Washington, D. C. May 15–18, 1967.
57. Haynes, B. W., and Hench, M. E., "Hospital Isolation System for Preventing Cross Contamination by Staphylococcal and Pseudomonas Organisms in Burn Wounds," *196 Annals of Surgery,* **162,** No. 4 (Oct. 1965).
58. Hellman, A., Personal Communication.
59. Hill, B. F. (ed.), "Proceedings of the Symposium on Research Animal Housing," *Laboratory Animal Care,* **13,** 221–467. (Symposium held Nov. 16–17, 1962, Washington, D.C.)
60. Hoffman, R. K., and Warshowsky, B., "Beta Propiolactone Vapor as a Disinfectant," *Appl. Microbiol,* **6,** 358–362 (1958).
61. Institute of Laboratory Animal Resources, National Academy of Sciences, National Research Council, "Guide for Laboratory Animal Facilities and Care," U.S. Department of Health Education and Welfare, Public Health Service, revised 1965.
62. Jemski, J. V., "Maintenance of Monkeys Experimentally Infected with Organisms Pathogenic for Man," *Proc. Animal Care Panel,* **12,** 89–98 (1962).
63. Jemski, J. V., and Phillips, G. B., "Microbiological Safety Equipment," *Laboratory Animal Care,* **13,** 2–12 (1963).
64. Kelley, S. M., Clark, M. E., and Coleman, M. B., "Demonstration of Infectious Agents in Sewage," *Am. J. Public Health,* **45,** 1438–1446 (1955).

65. Kethley, T. W., and Cown, W. B., "Dispersion of Airborne Bacteria in Clean Rooms," Proceedings, Fifth Annual Technical Meeting American Association for Contamination Control, Houston, Texas, March 30, 1966.
66. Kraft, L. M.; Pardy, R. F.; Pardy, D. A., and Zwickel, H., "Practical Control of Diarrheal Disease in a Commercial Mouse Colony," *J. Lab. Animal Care,* **14,** No. 1, pp. 16–19 (1964).
67. Kuster, E., "Die gewinnung, haltung und aufzucht keimfreier tiere und ihre bedeutung fur die erforschung naturlicher lebersvorgange," *Arb. Kaiserl, Gesundh,* **48,** 1–79 (1915).
68. Laboratory Animal Welfare, Federal Register, Volume 32, No. 37, Feb. 24, 1967, Washington, D. C. Part II. Dept. of Agriculture, Agricultural Research Service.
69. LeMunyan, C. D., Preliminary data presented at 16th Annual Meeting Animal Care Panel, Philadelphia, Pa., 1965.
70. Lennette, E. H., and Koprowski, H., "Human Infection with Venezuelan Equine Encephalomyelitis Virus. A Report of Eight Cases of Infection Acquired in the Laboratory," *J. Am. Med. Assoc.,* **123,** 1088–95 (1943).
71. Luckey, T. D., "Germ-free Life and Gnotobiology," Academic Press, New York, 1963 (512 pp.).
72. Machol, R. E., "Systems Engineering Handbook," McGraw-Hill, New York, 1965.
73. Mallis, A., "Handbook of Pest Control,"McNair Dorland Co., 1964.
74. Medical Diagnostic Laboratory Planning and Design (Proceedings of the Seminar on), Feb. 2, 1965, Baltimore, Md. USDHEW, PHS, CDC, Atlanta, Ga. 30333. Norman, J. C. "Air Handling Objectives," pp. 37–42.
75. Melnick, J. L., "Poliomyelitis Virus in Urban Sewage in Epidemic and in Nonepidemic Times," *Am. J. Hyg.,* **45,** 240 (March 1947).
76. Melnick, J. L.; Emmons, J.; Opton, E. M., and Coffey, J. H., "Coxsackie Viruses from Sewage. Methodology Including an Evaluation of the Grab Sample and Gauze Pad Collection Procedures," *Am. Hyg.,* **59,** 185–95 (1954).
77. Metcalf, C. T.; Flint, W. P., and Metcalf, R. L., Destructive and Useful Insects," McGraw-Hill, New York, 1962.
78. Michaelsen, G. S.; Ruschmeyer, O. R., and Vesley, D., "The Bacteriology of 'Clean Rooms,' " Final Report, Grant # NSG–643, from the National Aeronautics and Space Administration by School of Public Health, University of Minnesota, Minneapolis, Minnesota 55455, 1966.
79. Moore, B., "The Detection of Typhoid Carriers in Towns by Means of Sewage Examination," *Month. Bull. Min. Health and Pub. Health Lab. Serv.,* **9,** 72 (March 1950).
80. Morris, E. J., "A Survey of Safety Precautions in the Microbiological Laboratory," *J. Med. Lab. Technol,* **17,** 70–81 (1960).
81. Mosher, R. S., "Industrial Manipulators," *Scientific American,* **211**(4), 88–96 (Oct. 1964).

82. Munnecke, D. E.; Ludwig, R. A., and Sampson, R. E., "The Fungicidal Activity of Methyl Bromide," *Canadian J. Botany* **37,** 51–58 (1959).
83. National Institutes of Health, Bethesda, Md., "Planning and Design of Medical Research Facilities," 1962 (29 pp.)
84. National Research Council, Committee on Design, Construction and Equipping Laboratories, "Laboratory Design," Reinhold Book Corp., New York, 1951.
85. NIH Standard Animal Care Equipment. Available from: Procurement Section, Supply Management Branch, Office of Administrative Management, National Institutes of Health.
86. Norman, J. C., "The Bench Concept in Laboratory Planning," *Health Laboratory Science,* **1,** No. 3, July, American Public Health Association (1964).
87. Nuffield Foundation, Division for Architectural Studies, "The Design of Research Laboratories," Oxford University Press, London, 1961.
88. Nuttall, G. H. F., and Thierfelder, "Weitere untersuchungen uber bakterienfreie thiere," *Arch. Physiol.* (*Leipzeg*), pp. 363–364 (1896).
89. Perkins, J. J., "Principles and Methods of Sterilization," Charles C. Thomas, Springfield, Ill. 1956 (340 pp.)
90. Phillips, A. W., "Microbial Effects on Liver Monoamine Oxidase in the Chick," Abstr. 5th Intern. Congr. Nutrition, Washington, D.C., p. 26, 1960.
91. Phillips, C. R., "The Sterilizing Action of Gaseous Ethylene Oxide. II. Sterilization of Contaminated Objects with Ethylene Oxide and Related Compounds, *Am. J. Hyg.* **50,** 280–288 (1949).
92. Phillips, C. R., Gaserous Sterilization, pp. 746–765. *In* G. F. Reddish (ed.), Antiseptics, Disinfectants, and Chemical Fungicides and Physical Sterilization" (2nd ed.), Lea and Febiger, Philadelphia, Penna., 1957.
93. Phillips, G. B.; Novak, F. E., and Alg, R. L., "Portable Inexpensive Plastic Safety Hood for Bacteriologists," *Appl. Microbiol,* **3,** 216–217 (1955).
94. Phillips, G. B.; Jemski, J. V., and Brant, H. G., "Cross Infection Among Animals Challenged with *Bacillus anthracis," J. Infect. Dis.,* **99,** 222–26 (1956b).
95. Phillips, G. B., and Novak, F. E., "Applications of Germicidal Ultraviolet in Infectious Disease Laboratories. II. An Ultraviolet Pass-Through Chamber for Disinfecting Single Sheets of Paper," *Appl. Microbiol.,* **4,** 95–96 (1956).
96. Phillips, G. B.; Reitman, M.; Mullican, C. L., and Gardner, G. D., Jr., "Applications of Germicidal Ultraviolet in Infectious Disease Laboratories. III. The Use of Ultravolet Barriers on Animal Cage Racks," *Proc. Animal Care Panel,* pp. 235–244 (1957).
97. Phillips, G. B., and Hanel, E., Jr., "Use of Ultraviolet Radiation in Microbiological Laboratories," United States Library of Congress, P.B. 147 043. Listed in U.S. Govt. Res. Rept. 34(2), Aug. 19, 1960, p. 122.

98. Phillips, G. B., "Causal Factors in Microbiological Laboratory Accidents and Infections," University Microfilms Inc., Ann Arbor, Mich., 1965.

99. Phillips, G. B., "Microbiological Hazards in the Laboratory—Part One—Control," *J. Chem. Educ.,* **42,** A43–A48, "Part Two—Prevention," Ibid: A117–A130 (1965).

100. Phillips, G. B.; Edwards, R. W.; Favero, M. S.; Hoffman, R. K.; Lanahan, T. B.; MacLeod, N. H.; McDade, J. J.; and Skaliy, P., "Microbiological Contamination Control—A State of the Art Report," Biological Contamination Control Committee of the American Association for Contamination Control, April 1965.

101. Phillips, G. B., and Runkle, R. S., "Laboratory Design for Microbiological Safety," *Appl. Microbiol.,* **15,** No. 2, pp. 378–389 (1967).

102. Powers, E. M., "Microbial Profile of Laminar Flow Clean Rooms," Document X-600–65–308, Goddard Space Flight Center, Greenbelt, Md. (Sept. 1965).

103. Reddish, G. F. (ed.)., "Antiseptics, Disinfectants, Fungicides, and Chemical and Physical Sterilization." (2nd ed.), Lea and Febiger, Philadelphia, Penna., 1957 (975 pp.).

104. Reitman, M.; Frank, M. A., Sr.; Alg, R., and Wedum, A. G., "Infectious Hazards of the High Speed Blendor and Their Elimination by a New Design," *Appl. Microbiol.,* **1,** 14–17 (1953).

105. Reitman, M., and Wedum, A. G., "Microbiological Safety," *Public Health Rept.,* **71,** 659–665 (1956).

106. Reyniers, J. A., Introduction to the General Problem of Isolation and Elimination of Contamination, pp. 95–113, *In* J. A. Reyniers (ed.), "Micrurgical and Germfree Methods," Charles C. Thomas, Springfield, Ill., 1943.

107. Robertsen, J. A., and Payne, W. W., "The Professional Bio-hazard R & D Team," Presented at the Meeting of the Parental Drug Association, Chicago, Illinois, April 15, 1966.

108. Runkle, R. S., "Laboratory Animal Housing," *Am. Inst. Architects,* **41,** 55–58, 77–80 (1964).

109. Sadoff, H. L., and Almlof, J. W., "Testing of Filters for Phage Removal," *Ind. Eng. Chem.,* **48,** 2199–2203 (1956).

110. Schley, D. G.,; Hoffman, R. K., and Phillips, C. R., "Simple Improvised Chambers for Gas Sterilization with Ethylene Oxide," *Appl. Microbiol.,* **8,** 15–19 (1960).

111. Schreyer, J. M., and Bradford, C. M., "A Research Facility for the Study and Preparation of Ultrapure Materials," Document Y-DA-967, Union Carbide Corporation, Nuclear Division, Y-12 Plant Contract W-7405-eng-26, USAEC, Oakridge, Tenn., Sept. 17, 1965.

112. Schwartz, S.; Colvin, M.; Himmelsback, C. K., and Frei, E., "The Effect of Bacterial Suppression and Reverse Isolation on Intensive Chemotherapy," *Clinical Research,* p. 48 (Jan. 1965).

113. Seidler, F. M., "Adapting Nursing Procedures for Reverse Isolation," *The American Journal of Nursing,* **65,** No. 6 (June 1965).
114. Sell, J. C., and McMaster, W. W., "Planning the Electron Microscopy Suite," *J. Am. Inst. Architects* (May 1963).
115. Shepard, C. C.; May, C. W., and Topping, N. H., "A Protective Cabinet for Infectious Disease Laboratories," *J. Lab. Clin. Med.,* **30,** 112–716 (1945).
116. Slepushkin, A. N., "An Epidemiological Study of Laboratory Infections with Venezuelan Equine Encephalomyelitis," *Prob. Virol.,* **4,** 54–8 (1959).
117. Smith, V. C., "Distilled Water Distribution Systems." Available from: Barnstead Still and Sterilizer Co., Boston 31, Mass.
118. Snow, D. L., "Space Planning Principles for Biomedical Research Laboratories," Public Health Monograph 71, 1963.
119. Songer, J. R.; Sullivan, J. F., and Hurd, J. W., "Testing Air Filtering Systems. I. Procedure for Testing High-Efficiency Air Filters on Exhaust Systems," *Appl. Microbiol.,* **11,** 394–97 (1963).
120. Spiner, D. R., and Hoffman, R. K., "Method for Disinfecting Large Enclosures with B-propiolactone Vapor," *Appl. Microbiol.,* **8,** 152–155 (1960).
121. Sympósium, Sterilization by Ionizing Radiations, pp. 7–56. *In* "Sterilization of Surgical Materials," The Pharmaceutical Press, London, 1961.
122. Thorne, H. V., and Burrows, T. M., "Aerosol Sampling Methods for the Virus of Foot and Mouth Disease and the Measurement of Virus Penetration Through Aerosol Filter," *J. Hyg.,* **58,** 409–17 (1960).
123. Trexler, P. C., "Flexible-Wall Plastic Film Isolators," pp. 55–60. *In* Proc. 2nd Symp. Gnotobiotic Technology, 1959, Univ. Notre Dame Press, Notre Dame, Ind., 1960.
124. Trexler, P. C., "The Gnotobiote-Review and future," *Bio-Medical Purview* (Fall, 1961).
125. Tynsky, J. R., "Manual for Selection of Laboratory Benchtop Materials," (Project R-37) USDHEW, PHS, NIH, DRS, RFPB, June 1964.
126. U.S. Public Health Service, "Medical School Facilities, Planning Considerations and Architectural Guide," p. 75, Washington, D. C., Government Printing Office, 1961.
127. University of California Report of Statewide Animal Care Panel; Available from Institute of Laboratory Animal Resources, National Research Council.
128. Van den Ende, M., "Apparatus for the Safe Inoculation of Animals with Dangerous Pathogens," *J. Hyg.,* **43,** 189–194 (1943).
129. Van Duuren, B. L.; Orris, L., and Nelson, N., "Carcinogenicity of Epoxides, Lactones and Peroxy Compounds Part II," *J.N.C.I.,* **35–4,** 707–717 (1965).
130. Vesley, D.; Michaelsen, G. S., and Halbert, M. M., "Laminar Air Flow for the Care of Hospital Patients," *J.A.P.H.A.* (June 15, 1967).

131. Walburg, H. E.: Mynatt, E. I.; Cosgrave, E. G.; Tyndall, R. L., and Robie, D. M., "Microbiological Evaluation of an Isolation Facility for the Production of Specific-Pathogen-Free Mice," *Laboratory Animal Care,* **15,** No. 3.
132. Wedum, A. G., "Bacteriological Safety," *Am. J. Public Health,* **43,** 1428–1437 (1953).
133. Wedum, A. G.; Hanel, E., Jr.; and Phillips, G. B., "Ultraviolet Sterilization in Microbiological Laboratories," *Public Health Rept.,* **71,** 331–336 (1956).
134. Wedum, A. G., "Laboratory Safety in Research with Infectious Aerosols," *Public Health Repts.,* **79,** 619–633 (1964).
135. Wedum, A. G., and Phillips, G. B., "Criteria for Design of a Microbiological Research Laboratory," *A.S.H.R.A.E. J.,* **6,** 46–52 (1964).
136. Wolf, H. W.; Skaliy, P.; Hall, L. B.; Harris, M. M.; Decker, H. M.; Buchanan, L. M., and Dahlgre, C. M., "Sampling Microbiological Aerosols," Public Health Monograph No. 60, Public Inquiries Branch, U. S. Public Health Service, Washington 25, D. C., 1959 (53 pp.)
137. York, J. E., "Ventilation and Air Conditioning for Laboratories," *Heating and Ventilating's* Reference Section, Nov. 1953.

Appendix I

FEDERAL RADIATION PROTECTION REGULATIONS

Code of Federal Regulations, Title 10, Chapter I, Part 20.
Code of Federal Regulations, Title 10, Chapter I, Part 30.
Code of Federal Regulations, Title 49, Chapter I, Para. 72 and 73 (Transportation and Shipment).
NBS Handbook #42, "Safe Handling of Radioactive Isotopes."
NBS Handbook #48, "Control and Removal of Radioactive Contamination in Laboratories."
NBS Handbook #50, "X-ray Protection Design."
NBS Handbook #51, "Radiological Monitoring Methods and Instruments."
NBS Handbook #54, "Protection Against Radiations From Radium, Cobalt-60, and Cesium-137."
NBS Handbook #56, "Safe Handling of Cadavers Containing Radioactive Isotopes."
NBS Handbook #57, "Photographic Dosimetry of X- and Gamma Rays."
NBS Handbook #60, "X-ray Protection."
NBS Handbook #62, "Report of the International Commission on Radiological Units and Measurements (ICRU) 1956."
NBS Handbook #69, "Maximum Permissible Body Burdens and Maximum Permissible Concentrations of Radionuclides in Air and in Water for Occupational Exposure."
NBS Handbook #73, "Protection Against Radiations from Sealed Gamma Sources."
NBS Handbook #80, "A Manual of Radioactivity Procedures."
"Radiological Health Handbook," U. S. Department of Health, Education, and Welfare, Public Health Service (September 1960).

Appendix II

DESIGN CHECKLIST

Absolute Filters
Access to Waste Collection Treatment Rooms
Acoustical Ceiling
Air, Compressed
Air Conditioning
Air Flow Patterns
Air Incinerator
Air Locks
Alarm System and Fire Detection
Alarm System Biological Spill
Animal Holding Rooms
Autoclaves
Backflow Preventers
Barriers, Ultraviolet
Biological Safety Cabinets
Cabinets and Countertops
Cabinets—Flammable Solvents
Case Washers
Ceilings
Chemical Fume Hoods
Clean Receiving and Storage Rooms
Closed Circuit Television System
Closers, Door
Conductive-type Floor
Contaminated Laboratories
Contaminated Receiving Rooms
Contaminated Vent System
Corridor Widths
Curbs
Decontamination Equipment
Decontamination Showers
Deepfreeze
Deionized Water
Desks
Diaphragms, Speaking
Distilled Water
Doors, Air Locks
Drain Lines Size
Drainage, Air Lock Floors
Drains, Floor
Dunk Tanks
Emergency Exit Doors
Emergency Lighting
Emergency Power
Entrance, Service
Entrances, Personnel
Epoxy-aggregate Finish Floor
Exhaust System
Explosion-proof Areas
Finishes, Paint
Fire Sprinklers
Flammable Material Storage
Floor Covering
Foot-operated Valves
Fume Hoods
Gas, Propane
Gas Sterilizer
Gastight Doors
Heat Gains from Animals
High Vacuum Steam Sterilizer
Hood Areas
Hose Reels
Hospital Pullarms
Humidifier
Incubators
Inside Design Conditions
Intercommunications

Laboratory Equipment
Laboratory Furniture
Laboratory Sinks
Lighting Fixtures
Noncontaminated Receiving and Storage Rooms
Paint Finishes
Panels, Viewing
Partitions
Pass Box
Personnel Entrance
Protective Coating, Floors, Walls
Receiving Room, Contaminated
Receiving Room, Noncontaminated
Refrigerators
Refuse Incinerator
Rodent Proofing
Safety Showers
Sewage
Shoe Rack
Standby Electrical Service
Still
Storage Rooms
Suspended Ceilings
Telephone Systems
Traffic Control
Triple-distilled Water
Ultrasonic Washing Equipment
Ultraviolet Barriers
Vacuum
Vapor Barriers
Ventilated Cage Racks
Ventilated Suits
Walk-in Incubators
Walk-in Refrigerators
Washers, Glassware
Water, Deionized
Water, Distilled
Waterproofing
Windows

Index